POND BASICS

POND BASICS

Peter Robinson

Sterling Publishing Co., Inc.
New York

Creative Director	Keith Martin
Executive Editor	Julian Brown
Design Manager	Bryan Dunn
Senior Designer	David Godfrey
Designer	Les Needham
Editorial Manager	Jane Birch
Editor	Jo Richardson
Production Controller	Louise Hall
Picture Research	Wendy Gay
Illustrator	Damien Rochford

Library of Congress Cataloging-in-Publication Data
Available

10 9 8 7 6 5 4 3 2 1

Published by Sterling Publishing Company, Inc.
387 Park Avenue South, New York, NY 10016
First published in Great Britain by Hamlyn, a division
of Octopus Publishing Group Limited

Distributed in Canada by Sterling Publishing
c/o Canadian Manda Group, One Atlantic Avenue,
Suite 105, Toronto, Ontario, Canada M6K 3E7

Printed in China

Sterling ISBN 0-8069-2287-7

CONTENTS

INTRODUCTION

With its unique properties of reflection, sound and movement, water will bring a whole new dimension to your garden. It has an infinite capacity to both soothe and delight, whether it comes in the form of a small, restful pond or a dramatic fountain display.

In recent times, we have been brought to a re-evaluation of water as our most precious resource in the light of a growing worldwide shortage. A well-designed water garden offers the opportunity to appreciate and enjoy the many wonders of water without placing extra demands on valuable water reserves. More than that, creating a water feature makes a positive contribution to the natural world in attracting wildlife and aiding their survival in what is often an increasingly hostile environment.

There has never been a more interesting or easier time to create a water feature. Construction materials, accessories and plants are available in increasing variety, to bring water gardening within everyone's reach and to meet any individual's requirements to suit their own particular circumstances.

Pond Basics is a practical introduction to bringing water into your garden with the minimum of expertise, guiding you safely through the pitfalls to achieve success and satisfaction.

1 PLANNING

There is no substitute for water for sharpening the visual appreciation of light and form, and for creating mood. The architecture of surrounding plants and trees is reflected in the stillness of a pond, providing a focal point in the garden for quiet contemplation. In complete contrast, fountains bring vitality and a sense of drama with their captivating movement and rhythmic sound. As a calm water surface reflects form and structure, so moving water captures light. Streams become arteries of brightness in a lawn or rocky outcrop, while boulders and cobbles glisten in the splash of a waterfall.

choosing a water feature

Water offers such a wide variety of applications that no garden can be considered too small, no style too restrictive and few budgets too meagre to accommodate some kind of water feature.

There are two popular misconceptions to be dispelled at the outset in planning a water feature. Firstly, moving water needs no connection to the mains water supply. Secondly, ponds are not necessarily more appropriate in areas of high rainfall – they provide welcome oases in dry climatic zones.

When introducing water, spend time considering why you want it in your garden, then plan how it can best be integrated into its existing design scheme. Take into account the nature of movement and sound that moving water features can offer – these elements need to be applied with just as much care as colour and texture in a planting scheme. Adjustable flow controllers on the outlets of pumps to fountains and waterfalls enable the pressure and volume to be finely tuned – too much noise and it can be obtrusive; too little and it becomes irritating. Cobbles in shallow water at the base of waterfalls create a totally different sound to water falling into deeper water.

The considerations of site will be discussed in Excavation and Installation, page 22, but first let us look at some of the main types of pond.

FORMAL POOLS

A formal pool is often the ideal option for a small garden where there is little or no lawn and the surface is dominated by paving. The formal pool has clearly defined, crisp edges which are generally paved and form regular geometric shapes. The planting is restrained, confined mainly in aquatic planting containers, and dominated by specimen plants that have bold upright leaves, such as irises, which create a strong vertical contrast to the even, horizontal expanse of the water.

Left: The strong lines of this canal make a perfect frame for either a still mirror surface or fountain spouts in the small pools at either end. *Below right*: Rendering the wall arch and pool wall in the same colour focuses attention on the statue.

INFORMAL PONDS
The priority in planning an informal pond is to blend it into the existing style of garden design. This type of pond would have strong appeal to the plantsperson, offering considerable scope for lush planting which may not be possible in other parts of the garden. The boundaries of the pond could be extended to a bog area, so allow space for this at the planning stage. If fish are introduced, their choice, size and number will need strict limiting.

tip

- *Canals offer an ideal way of leading the eye to a focal point, for example an ornamental statue or urn.*

RAISED POOLS
These are similar to formal pools in their suitability for small spaces and being surrounded by paving. For small or patio gardens surrounded by high fences or walls, they introduce reflected light to counteract any feeling of restriction. One of the great pleasures of a raised pool is to be able to sit on the pool surround and enjoy the water at close quarters. Raised pools combined with a fountain are very suitable for fish because the fountain spray oxygenates the water in the summer months.

PONDS FOR FISH
If you aim to keep fish in a pond, it is important to resolve the practical issues of good fish husbandry at the initial planning stage in order to avoid problems when the fish are introduced. The serious fish-keeper's requirements are vastly different to those of the plantsperson. The design needs to allow for an increase in the amount of equipment such as pumps and

filters as the fish increase in size and number, as well as easy cleaning on a regular basis. Small fish grow into big fish, particularly where koi carp are concerned, and large koi are not the most suitable partners for ornamental water plants. Oxygenating plants that may be adequate for a small ornamental pond will not supply the needs of big fish.

WILDLIFE POOLS

These provide a constant source of interest, even in the winter months when the foliage dies back and the water surface lures bathing birds to the shallow edges. Providing these shallow bathing edges at a point that can be seen from the house is an important part of the planning process. A beach-like edge is not only important for bathing birds but also provides the vital means of entry and departure for amphibians. If the background to the garden is rural, wildlife pools help to blend the countryside into the garden.

CANALS

These are also suitable for formal settings. They use running water to form shallow, narrow streams which can include changes in height if the garden is sloping. On a level site, the canal forms a long, slender formal pool, generally without plants, which in addition to making a strong visual statement in a balanced garden design introduces a fine thread of reflected light.

FOUNTAIN POOLS

Fountains contribute the delightful sparkle of light and sound of water to pools. They can be used in both formal and informal pools but are most appropriate for formal ones.

Fountains display the full beauty of the light spectrum, making them the ideal centrepiece for formally designed gardens, particularly when there is also an ample clear water surface that reflects the symmetry of the surrounding garden and emphasize its parallel lines.

Fountain spouts can also add interest to walls, where a variety of self-contained or tiered fountains spilling into a base pool can be introduced.

Left: Using the same thin sandstone for both the edging and base stones makes this carefully crafted water course look more natural.
Below: Partially sinking a recycled barrel into the soil will give the swollen timbers more support and help the barrel to remain watertight.

STREAMS AND WATERFALLS

A slow, meandering stream will add enormous interest to a garden with the most modest of slopes and allow an expanse of lawn to be broken by movement and lush planting. Where there is sufficient slope, a faster-moving stream bordered by rocks offers the opportunity to plant alpines near the stream edge. Creative gardeners are in their element here, fashioning babbling sounds as water tumbles down rills and building small rock pools.

SELF-CONTAINED WATER FEATURES

Half barrels, old sinks and sandstone troughs are just a few of the containers that can bring water gardening into small spaces. In addition, there are an increasing number of self-contained kits that use tiny pumps to recirculate water through containers such as urns and cast-iron hand pumps. They can be easily incorporated into any style of garden, introducing sound and movement at the touch of a switch.

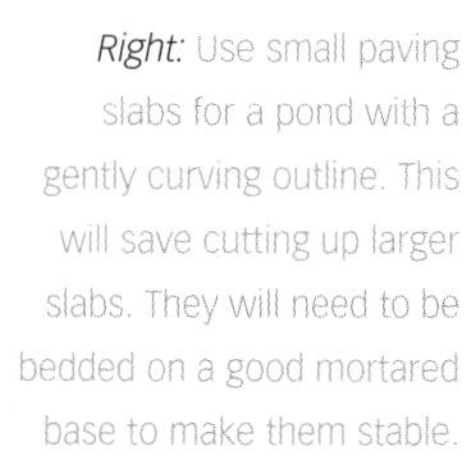

Right: Use small paving slabs for a pond with a gently curving outline. This will save cutting up larger slabs. They will need to be bedded on a good mortared base to make them stable.

safety measures

The question of whether it is safe to have a pond if young children use the garden poses a serious dilemma for parents or carers. No area of open water, no matter how shallow, can be considered entirely safe and a pool may have to be deferred for a while.

This postponement can be made easier and more interesting by creating a water-sympathetic feature in the meantime that could be developed to incorporate water at a later date. A dry scree bed leading into an area of cobbles would simulate a watery environment, which could subsequently become a stream and pond. Similarly, a sand pit could be built on the site of the future pond. Inexpensive polythene used to line a shallow excavation could be recycled as an underlay for the pool.

A more difficult situation is faced by a family with very young children that has moved to a new home where an existing pool makes an outstanding contribution to the garden. Draining the pool and filling it with sand may provide a temporary answer, but a more lasting, aesthetic solution may be to fence off the pool with an attractive picket fence and a lockable gate. In either case, the time will soon come when it is considered safe enough to fill the old pool with water or remove the fence yet some safety measures are still felt to be prudent. In this situation, there are a number of ways to make a pond a safer feature.

GRADING EDGES

Many pools in rural areas are remnants of old dew ponds, which are pools formed by puddling clay or digging out to the water table. These pools often have edges that fall sharply into the water, particularly in the summer months as the pool level sinks. This type of edge holds significant danger, with youngsters running towards the water and losing a foothold at the water's edge. This hazard can be reduced by regrading the edge to a gentle slope, so the water surface can be seen from some distance away rather than being hidden from view until the very last moment.

Gentle gradients around a pond leading to shallow, sloping submerged margins not only

Installing a protective grid

Covering the water surface with a strong protective grid will make the pond safer for young children.

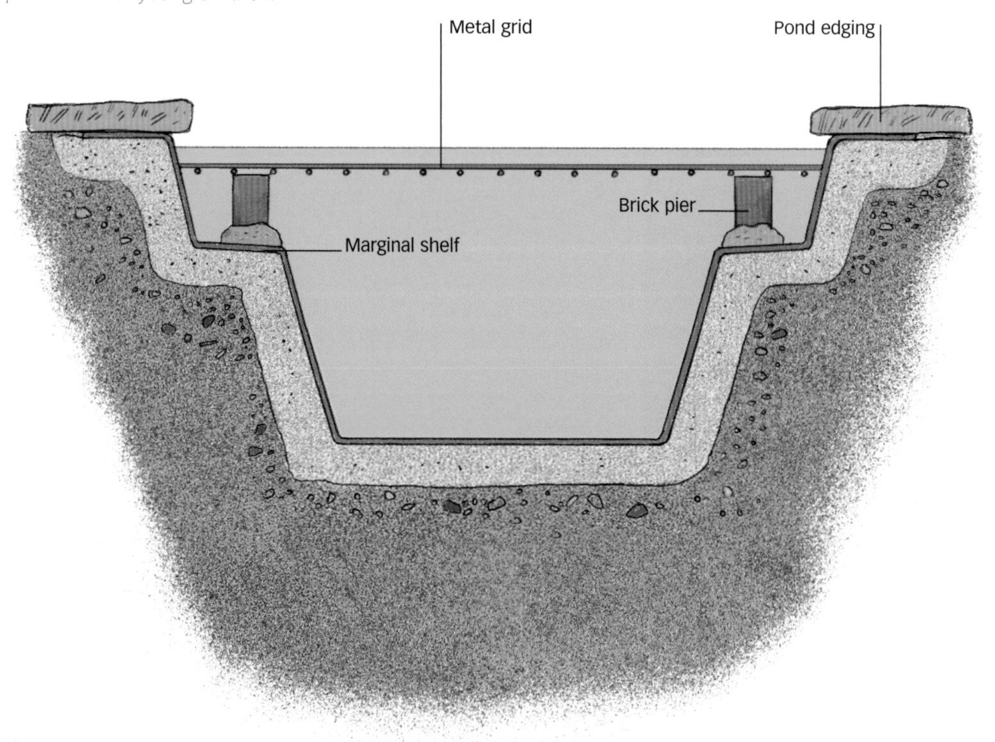

make it safer but more inviting, creating a greater sense of space. The pond will appear less deep and forbidding with no shadow at the water's edge. The design of any new informal pool surrounded by grass should allow for this restful grading for a short distance beyond the edges rather than a level area that suddenly changes to a steep edge at the side. Something else to avoid is moulding the soil excavated in the making of the pool around the immediate margins, thereby leaving a distinct hump. This not only looks unnatural but creates the danger of a steep slope immediately next to the water.

tip

- *An alternative way to conceal a protective grid if just below the water surface is to arrange cobbles on it in the same way as a cobblestone fountain – see page 53 for full instructions.*

SURFACING EDGES

The surface of any hard edge around a pool should be non-slip for all age groups. The most lethal edging is old, reclaimed stone paving slabs which attract algae and become slippery when wet. If natural stone must be used, use a stone with a riven surface for any paving slabs and keep them clear of algae. Many of the reconstituted concrete paving slabs are manufactured with roughened non-slip surfaces and these make good surrounds for the immediate edges of a formal pool.

Timber decking is now more widely used, but always ensure that it has a non-slip ribbed surface. Wet timber is just as

A cobblestone fountain is one of the safest types of water feature for a garden where small children play. They will also find it fascinating.

lethal as natural stone and planed surfaces are not intended for walking on. Decking tiles, which are laid with ribbed timbers in order to create a chequerboard effect, are also helpful in reducing the danger of slipping on wet, slimy wood.

USING PROTECTIVE GRIDS

A very small pond can be covered with a strong galvanized metal grid, which can then be disguised by growing several marginal plants through it. These grids are available at builders' merchants, where they are sold primarily for reinforcing concrete floors and paths. Their average size is approximately 2 x 1.2m (6½ x 4ft), so they they are adequate for a pond smaller than this where the grid will overlap the edges.

For larger ponds, a grid can be supported on piers inside the pool, enabling several panels of the grid to be butted together. If there are marginal shelves around the pool, the piers supporting the grid can be inside the pool, built on the marginal shelf, and need be no higher than two or three bricks. Longer piers would be necessary inside the deeper areas of the pool.

PLANT BARRIERS

A natural physical barrier can be created by thick planting in the pool margins and around the sides, to help prevent a fall into the water. This planting may limit the view of the pool while it is in place but can be moved when the children grow up. The vegetation is most effective if thick and woody at the outer edges up to the pond margins, where the sappier, softer leaves should then predominate.

Dogwoods (*Cornus*) and bushy willows (*Salix*) make good barriers where the soil is moist, but the range can be extended to other thicket-forming shrubs in drier soils. The design of wildlife pools, with their shallow beaches and thick vegetation around the sides, makes them good examples of relatively safe pools for gardens where children will be playing.

RAISED POOLS

Raised pools with side walls of 60–90cm (2–3ft) provide sufficient height for a small child not to reach the water. An additional measure to help prevent access over the side walls is to use coping slabs that are wider than the pool walls, to create an overhang of 5–10cm (2–4in) at the top of the walls.

The main danger of water to the elderly or disabled lies in inappropriate or badly laid edging which can cause slipping or tripping.

A raised or partially raised pool is an ideal solution to the problem of impaired vision or mobility, allowing the edge to become a casual seat.

Using low-voltage electricity

A low-voltage system can be used to operate a cobble fountain or pond lighting.

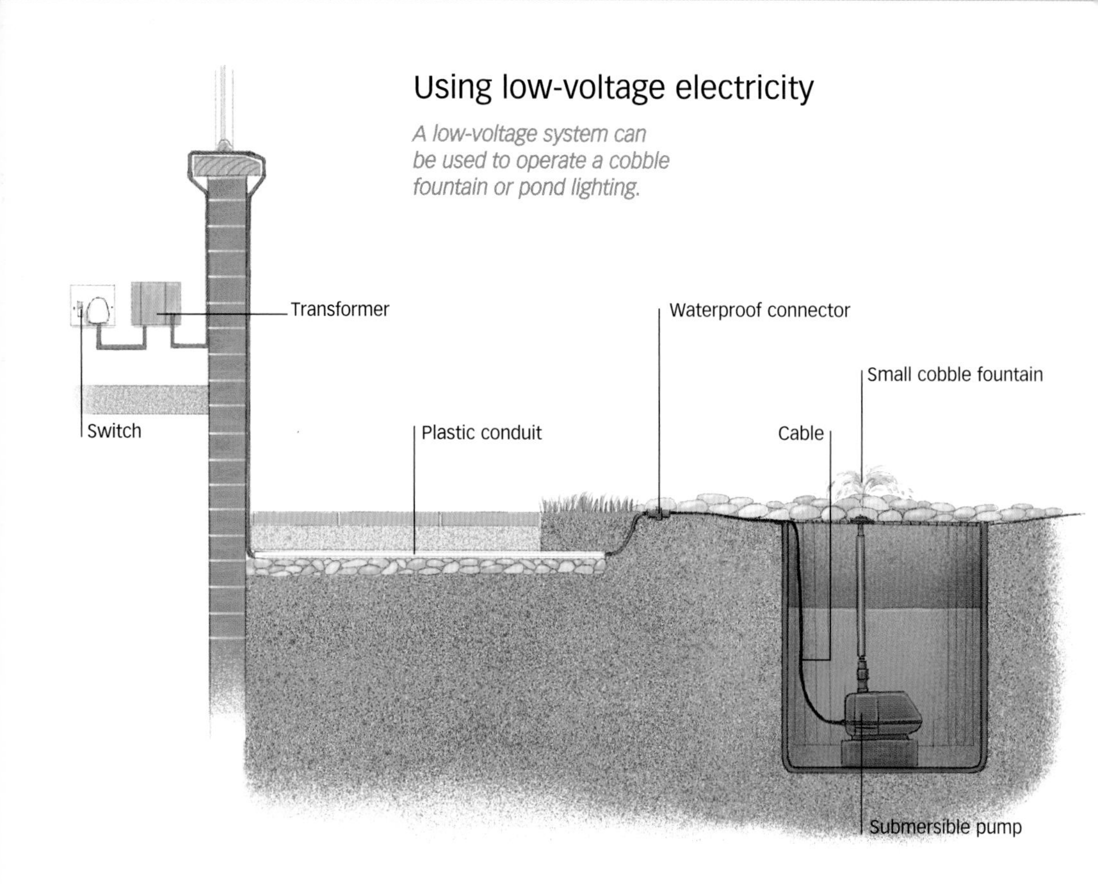

Using mains electricity

Main electricity is necessary to operate pumps for fountains over about 1m (3ft).

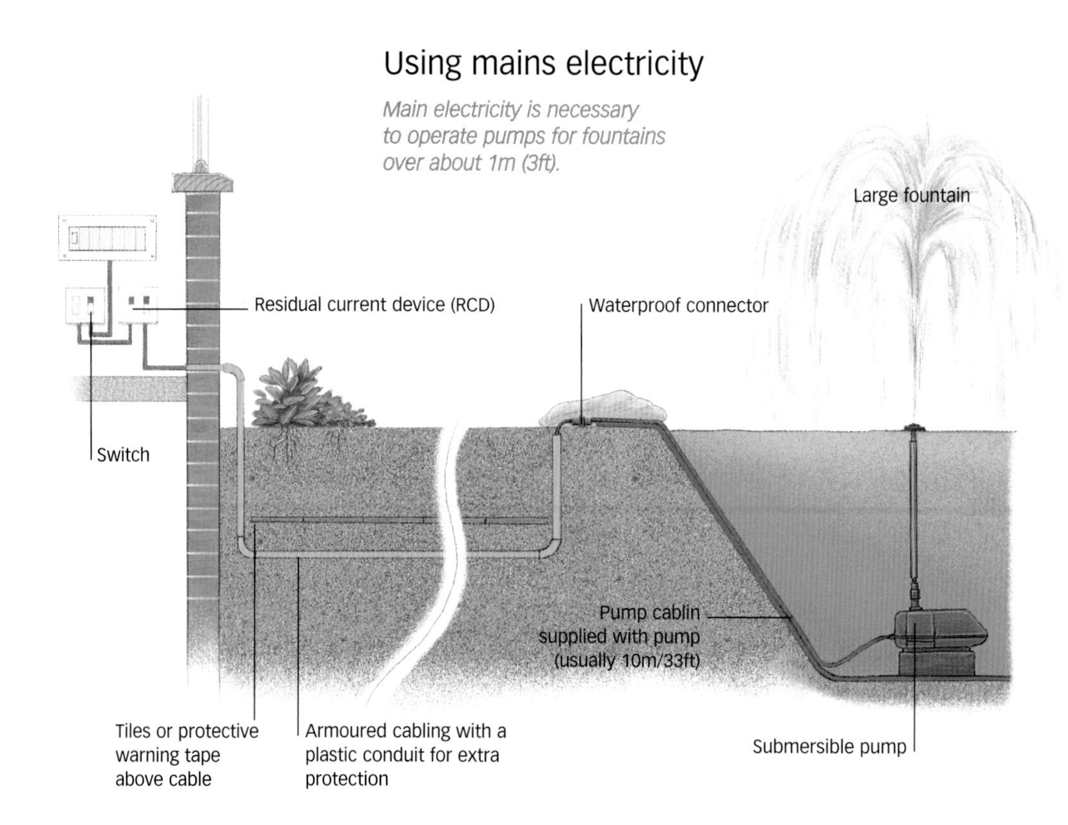

Right: Raised pools with walls built above toddler-height are less dangerous for children. The overhang makes climbing difficult and lets adults use the sides as casual seats.

PLAY FEATURES

A cobblestone fountain with a gurgling jet is a great source of fun for children as well as making an attractive garden feature. Construction is explained in detail on pages 48–9, and provided the supporting mesh for the cobbles is strong enough, there is no danger of falling in. A simple spray fountain that children can run through is a further refinement of the cobble fountain, in this instance the grid for the reservoir is fine enough to support a smaller grade of rounded pebble which it is easier to walk and run on.

ELECTRICITY AND WATER

There is an understandable reluctance to introduce cables carrying mains current to a pond. However, the risk of electrical shock has been almost totally eliminated by the application of residual current devices (RCDs), also known as contact circuit breakers, which are fitted to any electrical equipment outdoors where there is a danger of accidentally earthing mains voltage. These are extremely sensitive trip switches that are able to cut off the supply within 30 milliseconds of a possible leak or earth.

Although there are a number of electrical pond accessories, such as pond lighting, which are made safer by operating on low voltage using a transformer, the reduced voltage is not powerful enough to operate pumps for fountains over 1m (3ft) and the majority of these still operate on mains voltage. It is therefore advisable to note the following guidelines before installing electrical equipment.

- Use a qualified electrician for advice on both equipment and installation.
- Protect against accidental shock by installing an RCD at the connection to the mains voltage in the house, even if low-voltage accessories are used. If mains cabling is to be taken across the garden, use armoured cabling sunk to a depth of 30–60cm (1–2ft) under the soil and mark the position of the cable by covering it with roof tiles and warning tape.
- Use only approved waterproof junction boxes or switches to connect the mains cable to the cable supplied with the equipment. Position these where they are unlikely to be flooded.
- Keep a record of where any underground cables are located.

siting

Before taking the final step of choosing a site for the pond, there are some guidelines relating to its size, relative dimensions and profile that need to be considered.

SIZE

The general rule is the bigger, the better. This is because the larger the pool, the easier it is to manage as time goes on. In addition, most pool owners will admit to wishing they had built a larger pool because they do not have enough room to introduce new plants and/or their fish have outgrown the pool.

The smaller the pool, the more likely it is to have problems of green water, excessive temperature fluctuation and inadequate oxygen for the fish. If possible, try to achieve a minimum surface area of 4.5–5.5sq m (50–60sq ft).

PROPORTIONS AND PROFILE

Size is linked to depth. No matter how large the surface area is, if the pool is only 8–15cm (3–6in) deep, it will be a disaster. The ideal depth of medium-sized ponds with a surface area of 4.5sq m–18.5 sq m (50–200sq ft) is 60cm (2ft). Ponds less than 4.5sq m (50sq ft) in size may be 38–46cm (15–18in) deep. Ponds larger than 18.5sq m (200sq ft) would benefit from a depth of 76cm (30in). The reason for these guidelines is related to the needs of green algae. Algae thrive in warm shallow water in full sunshine. Deeper water allows for a greater volume of water that is not in the susceptible top 15cm (6in) where the algae thrive in the warmer and lighter conditions. The larger water volume also acts as a buffer to rapid and frequent temperature fluctuations, which are detrimental to many forms of pond life.

The relationship of depth to surface area is only valid when a pool has near to vertical side walls. In pools with a shallow, saucer-shaped profile, the volume can be reduced by as much as a half and an algae-free pool becomes more difficult to achieve. Smaller ponds with marginal shelves all the way round the sides also have a reduced volume, so it is better to restrict these shelves to where planting is necessary rather than build them to circumscribe the entire pool. Ornamental ponds need be no deeper than 76cm (30in) no matter how large the surface area.

OTHER FACTORS TO CONSIDER

With the concerns of size and profile in mind, the process of selecting the site for your water feature can begin. It might appear obvious to put a formal pool on the patio near the house or an informal pool at the lowest point of the garden, but to ensure that the best possible location is identified, the following points should be taken into account.

Shade

The pond should receive enough sunshine to warm the water and bathe the submerged plants in adequate light. The range of

Left: This formal pool is sited in the centre of the garden to catch as much sunshine as possible for the pool plants. *Below*: A deep area in a shallow cobble pool in a sunny spot helps to discourage algae and allows a clarifier and filter to be incorporated.

aquatic plants that can be grown in shaded pools is quite limited, and water lilies are reluctant to flower in these conditions.

Shade from trees is particularly troublesome because it is associated with leaf fall, which leads to a thick layer of decomposing vegetation on the pond bottom if not caught by netting (plastic mesh) placed over the surface. As this vegetation decomposes, methane gas is produced which is harmful to fish. It is especially important to prevent the leaves of yew, holly and laburnum sinking to the pond bottom since they are poisonous to other plant and animal life in the water. Although conifers may seem to be less of a problem, their leaves are constantly falling and depositing fine dusty bud scales on the water surface.

Wind

Wind cools the water surface, blows fountain spray and damages the soft succulent stems of marginal plants. In an attempt to capture the maximum amount of sun in a small garden, the pool is often sited in the centre of the lawn where it is more prone to wind exposure. Shelter can be provided by a trellis or planting on the pool's windward side. Any artificial windbreak should be semi-permeable rather than solid panels to prevent eddying and turbulence on the leeward side. Position the windbreak a little distance away from the pool, since the optimum effect at ground level of trellis or lath-like structures is felt at seven to ten times the height of the windbreak on the lee side.

Frost

Cold air accumulates in low-lying pockets, making plants more susceptible to browning by spring frosts. Informal ponds are often sited in low-lying areas because that is where they look more natural, making them more susceptible to frosts. The pond should be sited slightly higher up the dip in the land.

Sloping ground

A steep slope need not be too much of a constraint if the pond is partially dug out of the bank and partially banked up on the lower side. The outline of the pool needs to be slender and follow the contours of the slope. On land falling away from the house, build up the surround on the lower side of the pool so that the pool surface can be seen more easily. If the land rises from

Left: Small features like wall fountains liven up shady areas. Ferns make an ideal planting for the surrounds in shady places.

the house, cut more into the bank side so that the view is not spoilt by a high retaining wall.

A pond at the bottom of a slope is prone to frost damage and to run-off from the slope in heavy rain. If the surrounding grass has been treated with fertilizer which then leaches into the pond from the run-off, it can result in a sudden and dramatic growth of green algae in the pond as a consequence of the increase in nutrients. Drainage channels should be constructed where prolonged heavy rain or flash flooding is common.

Water table

The water table is the level at which water will stand in a ground hole or well. The level of the local water table rises and falls with the seasons and can be affected by large-scale building works or drainage schemes in the neighbourhood. Most water tables are well below the level at which a pond will be dug, but occasionally it may be a problem on wet, heavy land. A test to check if there is a high water table can be carried out by digging out a hole 60–90cm (2–3ft) deep and leaving it for a day or two to see if water appears in the hole. If water lies near the surface, there could be problems, since pond liners can billow up to the pool surface as a result of water pressure from the water table beneath.

Underground hazards

Once you have narrowed down your siting options, it is vital to ensure that a sunken pool will not be positioned over the route of underground services such as drains, gas pipes, water pipes, electricity cables and telephone cables. If there is any doubt over their route, contact the appropriate supply company, which will have the equipment to pinpoint the underground line.

tip

- *Sloping ground often has rocky substrata near the surface which may make excavation very difficult. Check that there is enough earth above the rock to allow for the depth of the pool by making small holes in the soil.*

FINE-TUNING THE SITE

The ideal site will probably be a compromise. For instance, high on your list of priorities may well be the ability to see the feature from a frequently used window of your home. Your choice may also be influenced by the desire to reflect any garden features such as ornaments or trees in the water. Sketching out a siting plan that identifies shade, prevailing wind direction, services and viewing lines from the windows is a useful step before finally taking out a hosepipe to lay on the ground and outline a shortlist of possible sites. Use a full-length mirror laid flat on the ground inside the pipe to simulate the effect of the water surface and see what reflections appear. If these are to be enjoyed from a particular window, take time making minor adjustments to the proposed position of the pond so that the view is at its best. A movement of 60–90cm (2–3ft) on the ground makes a great difference to the angle of reflection and how well the feature is framed.

Having identified the optimum site, decide how important an electricity supply is at the pool side. If a pump is vital and the distance makes this prohibitive in terms of cost, it may mean a final alteration to the site.

Where to site your pond

There are several factors to be taken into account when deciding where to site your pond.

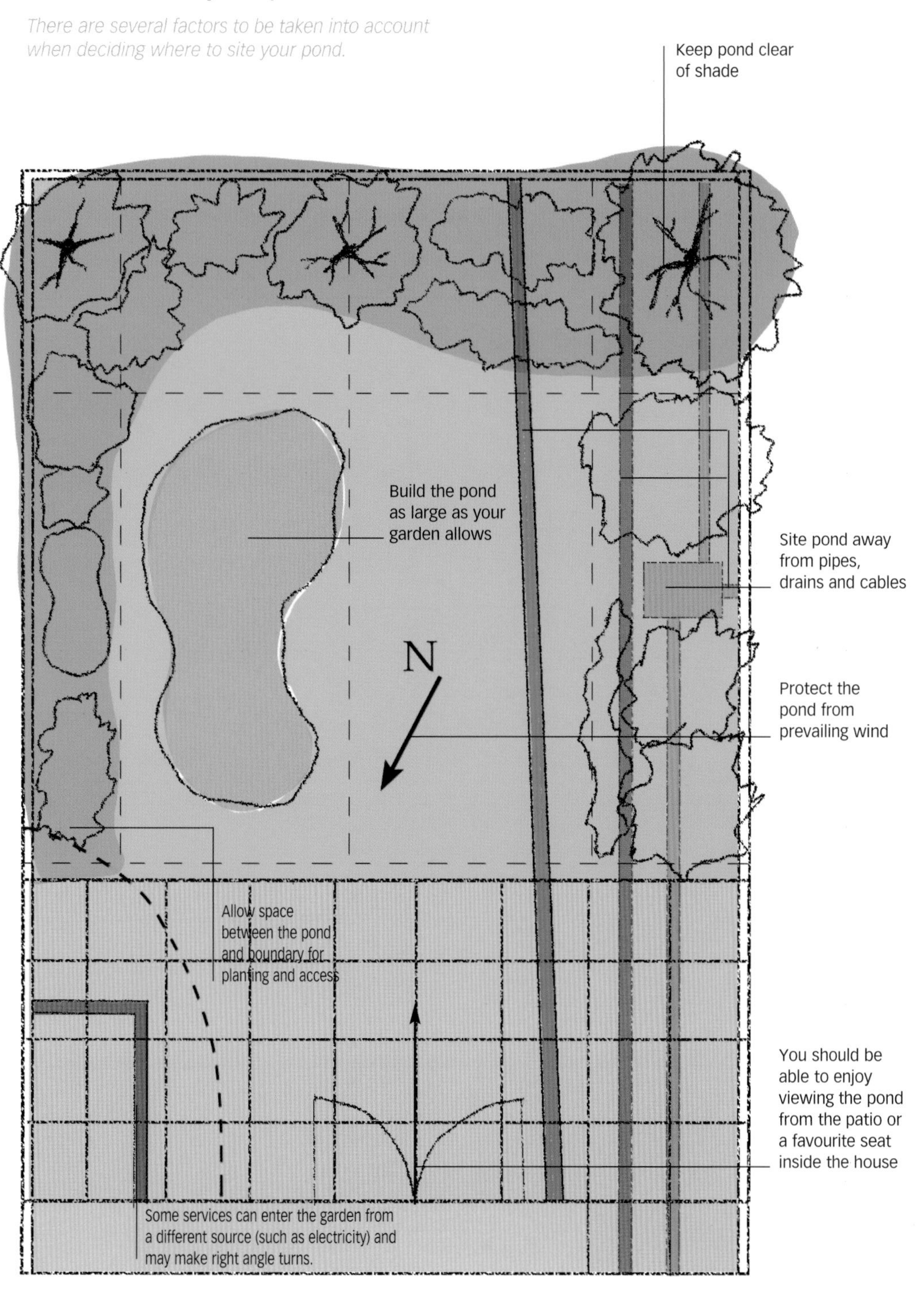

2 EXCAVATION & INSTALLATION

With a few exceptions, most of the jobs involved in pond construction can be successfully undertaken without previous experience in building. Furthermore, the variety of synthetic materials available for constructing ponds has never been greater or easier to obtain. From the wide choice of waterproofing membranes to the finishing touches for pond edges, there is ample choice to suit every design style. Once you have gained in confidence, the opportunity exists for you to extend a simple pond with a bed for moisture-loving plants, or perhaps a watercourse or a bog garden.

excavating techniques

When excavating in an area of lawn or rough turf, the top layer of turf should be removed first and, if space is available elsewhere in the garden, the turves stacked upside down in a neat pile. After a few months rotting down they will produce a fibrous loam which is ideal for potting aquatic plants. The next layer, 30–38cm (12–15in) deep, is topsoil, which is also valuable as a potting medium for aquatics or a top-dressing on the borders. This soil will prove invaluable if some slight alterations need to be made to the contours around the pool once it has been installed. The bottom layer, which is generally a different colour to the topsoil, is subsoil and should be discarded unless significant changes in level are anticipated.

A mound can be created near the pool to make a base for a watercourse, but if on a flat site, make sure it does not look too contrived by keeping the proportions and gradient as natural- looking as possible. The height of the mound should be no more than one fifth of the width, with gently sloping sides.

WATER SEEPAGE

The only unforeseen problem that very rarely occurs in excavation is water seeping into the hole on wet, heavy land, particularly in winter. The level of the water table on wet soil should be checked beforehand (see page 20), but where there is still a problem, one of the options below will remedy the situation:

- Raise the proposed level around the pond by adding extra soil.
- Move the pond site to higher ground if available.
- Build a raised or semi-raised pool on the site (see pages 34–5).
- Try to drain the area around the proposed pond.
- Pump the water out of the excavation, then use a heavy fibreglass pond unit (see page 24) weighted down inside with bricks or concrete blocks.

A drain plug is not necessary in the bottom of ornamental ponds; there is always the risk that it may not be 100 per cent watertight. Such a plug is only necessary in fish ponds where stricter hygiene and regular cleaning out are required.

preformed ponds

A preformed pond is the ideal choice for creating a pool with a symmetrical shape. There are two types of preformed pond – the rigid forms, which are made of fibreglass or reinforced plastic, and the thinner semi-rigid forms, made from a cheaper plastic which is moulded into sophisticated shapes under a vacuum process. Both types are more suitable when prefabricated into simple shapes rather than having over-fussy or narrow outlines. Square, rectangular and circular shapes make it easier to pave around the edge and disguise the unit. The stronger units are easy to clean out and repair if necessary, and their shiny surfaces make it a simple task to remove algae.

Rigid and semi-rigid pools are relatively easy to install compared to installing a flexible liner, which involves folding in tight corners. Rigid pools are also very useful on a sloping site where one end of the unit is rigid enough to be partially raised and disguised with either soil or a retaining wall.

Most rigid pools have preformed shelves around the sides which enable aquatic planting baskets to be housed in the shallower water. When the containers are closely packed together on the shelves, the plants blend with each other, giving an informal and established appearance in a short period of time.

In general, preformed ponds are less appropriate for informal pools, particularly those that have irregular, slender outlines and several shelves at different

Left: Preformed ponds are better than flexible liner for regular shapes like circles, squares and rectangles, as there is no problem with awkward folds. The rim can be disguised with a paving surround.

heights which reduce the overall volume, making the water prone to greening. It is vital to ensure they have a minimum depth of 46cm (18in); units shallower than this will be vulnerable to rapid temperature changes and unsuitable for very hot summers or cold winters if fish are present.

Preformed ponds become extremely heavy when full and are subjected to considerable water pressure if the excavation does not support their shape evenly. This can cause hairline cracks in the cheaper units which are difficult to detect once they are filled with water. Bear in mind that preforms look deceptively large when displayed on their sides at the time of purchase and much smaller when sunk into the ground. The fact that they can be easily handled and transported puts limitations on their size.

EXCAVATION

1 In order to mark out the site for a symmetrically shaped preformed pool, simply invert the unit onto the proposed site and outline with sand around the edge of the rim.

2 For asymmetrical shapes, stand the unit upright on the proposed site and temporarily support it with bricks or walling blocks to prevent it from falling over. Push canes from the rim into the soil directly beneath the unit at about 1m (3ft) intervals. Run a length of string around the canes to mark out the outline, which can then be marked out using sand.

3 Measure the depth of the unit from the rim to the bottom of the marginal shelf. Remove the soil down to this depth from 5–8cm (2–3in) outside the outline.

4 Lightly rake and level the dug surface. Place the unit in the prepared hole and press it down firmly onto the raked surface so that an impression is made of the base. Lift out the unit, then dig from 5–8cm (2–3in) outside the area marked by the unit base.

5 When the depth of the unit plus an extra 5–8cm (2–3in) for a layer of soft sand has been reached, lay the straight-edged length of wood across the width of the hole and with a tape measure check that the hole depth is correct. Ensure that the base of the hole is level using a spirit level.

6 Rake over the base and sides to remove tree roots and sharp stones, and firm the base evenly before spreading the layer of sand across the bottom.

INSTALLATION

1 Enlist help to lower the unit into the hole. Check that it is sitting level by laying the straight-edged length of wood across the unit sides and using a spirit level.

2 Gently pour water into the unit to a depth of 10cm (4in), then backfill with sand or sifted soil between the sides of the unit and the sides of the hole to the same depth as the water. Use the flat-ended length of wood to ram down the backfilling. This process is known as tamping.

3 Repeat this procedure by adding a further 10cm (4in) depth of water to the unit, then 10cm (4in) of sand or sifted soil around the sides at a time, ensuring that no air pockets are left in the backfilling and the unit continues to remain dead level, until the pool is nearly full and the weight of the water will keep it stable. After this the edging can be put in place.

INSTALLING A PREFORMED POND

You will need

- *Sand, bamboo canes and string for marking out*
- *Preformed pool unit*
- *Bricks or walling blocks for temporarily supporting an asymmetrically shaped unit*
- *Tape measure*
- *Spade*
- *Wheelbarrow*
- *Rake*
- *Soft sand or sifted soil*
- *Straight-edged length of wood, long enough to straddle the sides of the unit*
- *Spirit level*
- *Flat-ended length of wood, 5 x 5 x 60cm (2 x 2 x 24in), for tamping the soil*
- *Suitable edging materials (see pages 55–9)*

Left: Flexible liners are ideal for larger and more individual ponds because they allow more freedom of shape.

ponds with a flexible liner

Flexible liners provide the greatest scope for different shapes and designs of pool. They can be used entirely on their own to create informal ponds on heavy soils, or in combination with concrete or walling blocks to secure the sides of the excavation for formal pools.

Flexible liners are available in a variety of materials, colours and thicknesses, sold from rolls of varying widths or welded to specific sizes for larger applications. The choice of material will depend on available resources and the style of the pool. The most expensive is butyl and the cheapest is polythene, and between these two types there is a range of excellent materials suitable for most applications. Reducing the size of the pool to buy the most expensive liner is not recommended. If the liner is to be covered with soil or other materials, such as cobbles, the more inexpensive varieties are perfectly adequate since they will not deteriorate through exposure to ultraviolet light.

Where different colours are available, choose black – it looks more natural and gives a greater illusion of depth.

TYPES OF FLEXIBLE LINER

POLYTHENE (POLYETHYLENE)

- The oldest type of pool liner, developed in the 1930s
- Available in different thicknesses (only the thickest grades are suitable for lining ponds) and roll widths
- Cheap

Disadvantages:

- Deteriorates in ultraviolet light by hardening or cracking, making it the least durable liner
- Unwieldy to handle
- Easily torn
- Separate pieces cannot be joined
- Cannot be repaired
- Life-expectancy of 3–5 years
- No guarantees usually given

LOW-DENSITY POLYTHENE

- A recently developed and improved form of polythene that is becoming more widely available
- More flexible than standard polythene
- Difficult to tear
- Can be repaired
- Extremely slow deterioration in ultraviolet light, giving a life- expectancy of 15–30 years, depending on the strength of surface coating applied in the manufacturing process
- Guarantees available for 15–30 years
- Cheaper than alternative liners

Disadvantages:

- Separate pieces cannot be joined

PVC

- Developed from a new generation of polymers in the 1960s
- Available in different thicknesses and densities
- Certain grades laminated and reinforced with nylon netting for extra strength
- Separate pieces can be joined
- Can be repaired
- Longer durability than ordinary polythene with a minimum life-expectancy of 15–20 years
- Most grades supplied with guarantees of different lengths
- Variable in cost

Disadvantages:

- Heavy-duty types are not very malleable or flexible

BUTYL

- Developed about the same time as PVC and still remains the most commonly used liner for the professional installer of water gardens
- Available in various thicknesses
- Its elasticity makes it unique, giving it more strength than other liners and allowing it to fit into awkward shapes more easily
- Easily extended by welding, which can be done on site if required
- Easily repaired; being a rubber product, cycle inner-tube repair kits can be used
- Resistant to ultraviolet light
- Long life-expectancy – at least 50 years.
- Most suppliers give a guarantee of a minimum of 20 years

Disadvantages:

- Costly

UNDERLAYS

Protective underlays for ponds are now widely available in the form of rolls of non-woven geotextiles which are impenetrable even by the sharpest of stones. These have replaced sand or newspaper, sand being unstable on the sides of the excavation and newspaper eventually rotting under the liner and exposing it to sharp objects as the weight of the pool settles the soil underneath.

Constructing a pond using a flexible liner

Using flexible liner allows you the scope to create a wide variety of pond shapes.

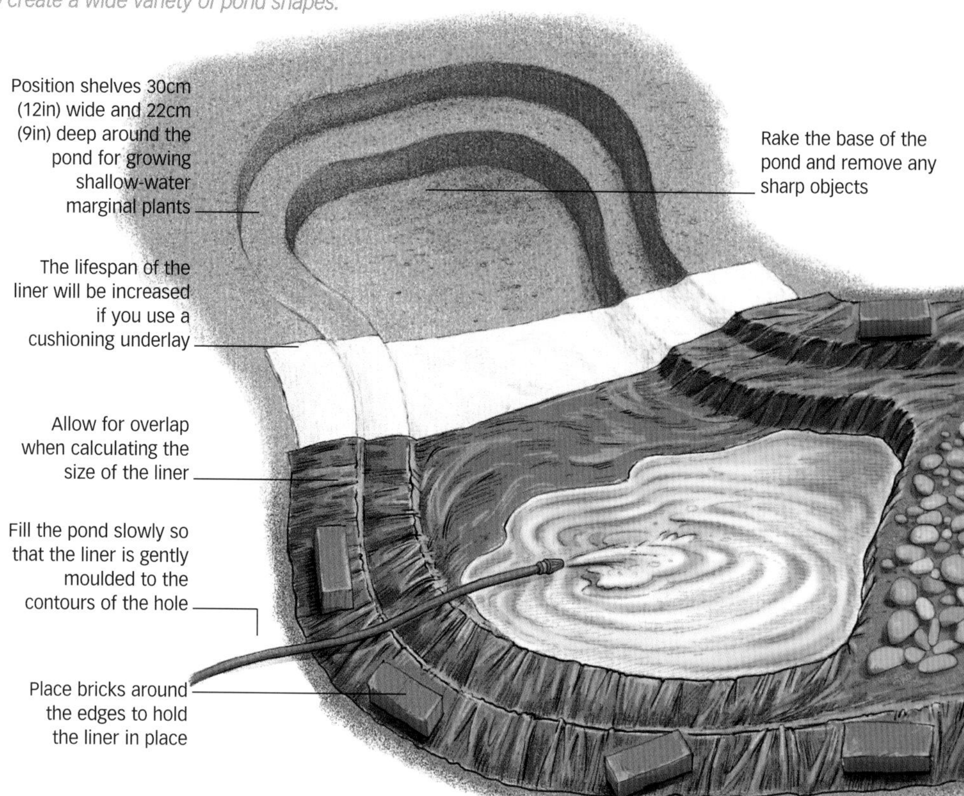

USING FLEXIBLE LINERS

Flexible liners offer the maximum amount of flexibility in the construction of ponds and other water features. They are now so universally obtainable that there is a good case for completing all the necessary excavations before purchasing the liner. By completing the digging first, there are no limitations on making last-minute adjustments to the shape and depth of the pool. After the excavation has been completed, move well away from the site and view it from as many angles as possible to see if any adjustments would be an improvement.

One rectangle of liner can provide a variety of pond shapes, including designs with narrow waists to make a crossing point for added interest. Where the wastage of liner would be excessive for very narrow sections, smaller pieces can be welded together at specialist suppliers or taped together on site using proprietary waterproof joining tape. While large creases in the corners of rectangular pools or sharp curves in informal shapes are inevitable, they can be made to look less conspicuous if the liner is carefully folded before the pool is filled. With time, the covering of algae and submerged planting will disguise the folds of the liner even further.

By allowing an ample amount of surplus liner around the sides, additional features such as a bog garden can be made. When creating a kidney-shaped pool, for instance, a small bog area can be made by using the corner piece of a rectangular sheet of liner. Instead of cutting

tip

- *Flexible liners are invaluable for waterproofing old concrete pools. These pools are notorious for sprouting fresh hairline cracks each year despite receiving regular attention.*

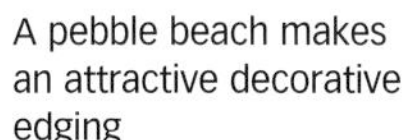
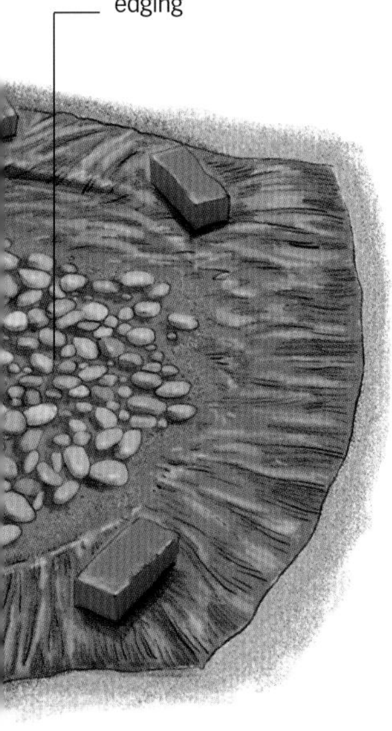

off the surplus liner, soil can be placed on the liner and prevented from spreading into the water of the main pool by a small submerged retaining wall of rocks or walling stones.

CALCULATING THE SIZE OF THE LINER

Measure out a rectangle that will enclose the outline of the pool. Measure the maximum length and breadth of the pool, then add twice the depth of the pool to each measurement. This is the bare minimum of liner required, so add 30cm (1ft) to each measurement to provide a small overlap. For brimming pools, add a little more than the width of the paving or bricks that surround the pool so that the liner can extend beneath and behind the edging to create a vertical lip.

BUILDING A POND WITH A FLEXIBLE LINER

You will need

- *Hosepipe, string or sand for marking out*
- *Half-moon edging tool (if pond is sited in a lawn)*
- *Spade*
- *Wheelbarrow*
- *Wooden pegs approximately 2.5cm (1in) in diameter and 15cm (6in) in length*
- *Heavy hammer*
- *Spirit level*
- *Straight-edged length of wood approximately 2m (6½ft) long*
- *Large polythene sheet*
- *Rake*
- *Liner*
- *Underlay*
- *Bricks or heavy stones*
- *Suitable edging materials (see pages 55–9)*
- *Large scissors*

EXCAVATION

1 Using a hosepipe, string or sand, mark out the pond outline.

2 If the pond is to be sited in a lawn, remove the turf by stripping off the grass to a depth of 2.5cm (1in) in squares of 30cm (12in) and stack upside down in another part of the garden for later use. Use a half-moon edging tool to make a neat edge to the turf.

3 Knock in the pegs at 2m (6½ft) intervals 15cm (6in) outside the outline. The pegs need to be sufficiently close so that the straight-edged length of wood can straddle the adjacent pegs.

4 Make a mark on one peg at soil level that is at the desired level of the finished pond. This is known as the datum point. Mark the other pegs to match, so that any variations in level around the outline can be seen and adjusted by adding or removing topsoil.

5 Using the spirit level and the length of wood, adjust the depth of the pegs so that their tops are all level.

6 Begin digging out the hole 15cm (6in) inside the pegs to a depth of 23cm (9in), angling the sides inwards slightly. This reduces the risk of damage by expanding ice in severe winters and of the walls subsiding. Store the soil on the polythene sheet nearby if it is likely to be used for contouring the pond surrounds later on.

7 Rake the base of the hole, then use sand to mark the position of any marginal shelves around the perimeter of the hole. These shelves should be 30cm (12in) wide and positioned where you want shallow-water plants.

8 Dig out the inner or deeper zone, avoiding the marginal shelf outlines, to the full depth of the pond – a further 23cm (9in) if the pond is to be 46cm (18in) deep or a further 38cm (15in) if 60cm (24in) deep.

9 Rake the bottom of the pond to level the surface.

10 Gently rake over the sides and bottom to remove any sharp or protruding surfaces.

Left: Rocks and water are natural partners. Here, rocks have been placed to allow rivulets of water to form. *Right above:* Mark out and remove the turf to the shape of the pool. Then excavate the pool to the required depth. *Right below*: Build a small concrete foundation on the marginal shelf and then apply a layer of mortar before draping the liner into place. *Far right above*: With the liner installed, lay a course of walling stones onto it and secure this with mortar. *Far right below*: Roll or lay the turf back onto the walling stones and finish the rock edge around the rest of the pool.

INSTALLING A FLEXIBLE LINER

1 If using a small-sized liner, lay it out on any lawn area adjacent to the excavation to warm up in the sun, thereby smoothing out any creases and making it more pliable to work with.

2 Drape the underlay across the hole and shelves, leaving an overlap of approximately 30cm (12in) all round the pool.

3 If using a large-sized liner and it remains folded, follow the supplier's instructions for putting it into position. This procedure involves placing the folded liner in the centre or at the end of the hole and unrolling until it is loosely draped across the excavation. If the liner is small enough and help is at hand, lift it by the corners and lower it into place, taking great care not to disturb the underlay as you do so.

4 Creases inevitably appear in the liner as it is draped into irregularly shaped holes, but try to make a few larger folds rather than several smaller ones.

5 Place bricks or heavy stones around the edges to hold the liner in place and prevent any wind from blowing the edges into the pool.

6 Before introducing water, checking that there is plenty of liner above the pegs all the way around the pond.

7 Fill with water to within a few centimetres of the tops of the pegs so that any final adjustments to the level of the sides can be made. Remove the bricks or stones holding the edges of the liner.

8 Build or add any edging before filling with water to the final level. Trim the surplus liner with scissors only after you are absolutely satisfied that the water and edging is satisfactory.

VARIATIONS

For an informal garden, a refinement of the basic pond with a flexible liner is to build a rock edge and grow marginal plants in the wet soil behind the rocks, making use of the soil removed from the pond excavation. It looks more natural to leave a portion of the edge as lawn rather than surrounding the whole pond with rocks. However, since the grass edge will probably have to cope with extra wear as a viewing point for the pool, it is given a more solid foundation with a small supporting wall, which will be partly underwater when the pool is filled. This requires a basic degree of bricklaying skill.

EXCAVATION

Follow Steps 1–10 of the excavation procedure on page 29 but in Step 7 make the marginal shelf approximately 46cm (18in) wide to accommodate the width of a piece of rock and enough soil for the marginal plants.

INSTALLATION

1 Roll the turf back by 30cm (12in) along one side of the pond and leave it rolled up in place.

2 Before draping the liner across the excavation, make a small foundation of concrete 10–13cm (4–5in) thick and 15cm (6in) wide along the section of marginal shelf corresponding to the section of turf edging.

3 Follow Steps 1–6 of the installation procedure given on page 28.

4 With the liner in place, lay one course of a walling stone onto a bed of mortar over the position of the concrete foundation. The reinforcing wall should finish 2.5cm (1in) beneath the level of the grass.

5 Once the mortar has set, backfill any remaining gap between the reinforcing wall and the liner with tamped soil before rolling the turf back to rest on top of the wall.

6 Lay rocks onto the remainder of the marginal shelf. For extra stability, lay them on a base of stiff mortar which is 5–7cm (2–3in) deep.

7 Once the mortar has set, backfill between the rocks and the liner with soil.

8 Position oxygenating plants and water lilies on the bottom of the pond before completely filling the pond.

9 Bring the water up to its final level and plant marginal plants in the soil between the rocks and the liner. Roll back the remaining rolled-up turf to the pool edge.

Left: As well as being ornamental, the lush vegetation growing from boggy soil provides ideal habitat and cover for a rich variety of wildlife.
Right: Careful choice of plants means that even boggy shaded areas need not be entirely green.
Far right: The slender spiked and diminutive size of *Typha minima* make it ideal for smaller bogy areas.

Wildlife pool with adjacent bog garden

Adjacent bog gardens are an essential part of a wildlife pool, and can make use of a portion of the flexible liner used for the pool itself. They provide both extensive cover and homes for many of the amphibians that hibernate out of the water but stay near the pool.

EXCAVATION AND INSTALLATION

1 Dig a hole for the pond following Steps 1–10 of the excavation procedure on page 29, but do not include any marginal shelves and make the pool saucer-like in profile with shallow sloping sides. As an extra precaution on soils containing sharp stones or flints, line the excavation with a 5cm (2in) layer of sand.

2 Mark out the overall site for the bog garden adjacent to the pool excavation. At intervals within this area, dig holes 0.9–1.2m (3–4ft) in diameter and 30–38cm (12–15cm) deep to

Left: Different bog plants need different depths of soil for their roots and different amounts of room to spread into. Dig out the various areas before lining the pool and bog. The area allocated to bog plants often exceeds the surface area of the pool

contain larger native marginal plants, such as bulrush (*Schoenoplectus lacustris* subsp. *tabernaemontani* 'Zebrinus'), sweet flag (*Acorus calamus*) and yellow or flag iris (*Iris pseudacorus*). Intersperse smaller depressions 60cm (2ft) in diameter and 15–23cm (6–9in) deep for smaller marginals such as marsh marigolds (*Caltha palustris)* and water forget-me-nots (*Myosotis scorpioides*).

Using a straight-edged length of wood and spirit level, check that the height of the sides of the depressions are no higher than the sides of the pool, so that the pond water will overflow from the main body of the pond into the depressions.

3 Drape the underlay followed by the liner over the pond and bog garden area.

4 Fill all the bog garden depressions with soil. The outline of the sides of the depressions will be visible as the soil settles or is watered. Add a covering of soil 10–15cm (4–6in) deep over the bottom of the pool.

5 Use the back of a rake to level off the bog garden area.

6 Lay cobbles around the section of pond edge opposite to the bog garden area, starting just under the surface to just above the surface, forming a beach (see pages 58–9). The shallow sides will prevent the cobbles from rolling into the pond. If the gradient is too steep for the cobbles to remain stable, attach them to the liner with a layer of mortar. Fill the pond with water.

7 Plant the various marginal plants in the appropriate soil-filled depressions in the bog garden, using a plank to gain access on the wet soil.

Left: This raised pool echoes the formal lines of the patio and provides a focal point.

Raised and semi-raised ponds

Raised and semi-raised pools make excellent focal points in formal patio or courtyard gardens, and provide an ideal context for fountains. They are easier to empty than sunken pools, suffer less from the problem of leaves and other wind-blown plant debris, require very little excavation and are an ideal solution for sloping sites. To stand up to the internal water pressure, raised pools with a flexible liner must have substantial surrounds, such as twin walls, making them more costly to build than sunken pools, and if built with brick or walling stone, some degree of bricklaying skill will be necessary.

RAISED POND WITH LOG ROLL SURROUND

A simple raised pool can be made by partially installing a rigid preformed unit and surrounding the outline with log roll, which is also flexible enough to curve with the contours of an informal pool. Log roll consists of machine-rounded semi-circular logs 7.5cm (3in) in diameter, fastened together in a form of rail or fence with galvanized wire for durability. Log roll is available in natural green or brown finish in 1m (3ft) lengths and heights of 10, 20, 30 or 40cm (4, 8, 12 or 16in). The height of log roll required is determined by the depth at which the rigid unit is sunk into the ground and so how far it projects above ground.

Sinking the deeper zone of the unit only, leaving the marginal shelves above ground level, allows for reasonable frost protection in winter and cool water in summer. The length of log roll required is calculated by placing a tape measure loosely around the unit.

INSTALLATION

1 Clear the ground underneath the site to expose the soil, and rake to produce a shallow tilth. Lift the unit onto the raked soil and press down to make an imprint of the deeper zone.

2 Put the unit to one side and dig from 5cm (2in) outside the marked area to a depth of 5cm (2in) more than the deeper zone. Rake the bottom level and add a 5cm (2in) layer of soft sand.

3 Lift the unit into the hole and use the straight-edged length of

Cross-section of a raised pond

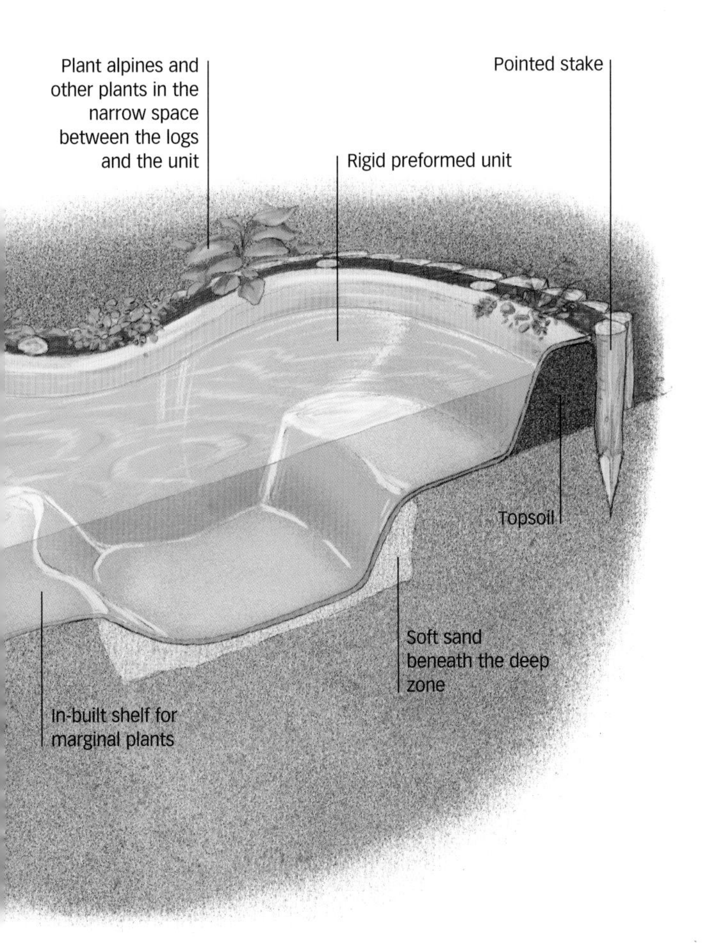

CONSTRUCTING A RAISED POND

You will need

- *Rigid preformed unit*
- *Spade*
- *Rake*
- *Soft sand*
- *Straight-edged length of wood*
- *Spirit level*
- *Treated pointed stakes 46 x 4cm (18 x 1½in)*
- *Heavy hammer*
- *Log roll 20cm (8in) high, long enough to surround the unit*
- *Wire cutters*
- *Galvanized nails*
- *Sifted topsoil*
- *Flat-ended length of wood, 5 x 5 x 60cm (2 x 2 x 24in), for tamping*
- *Selection of alpine and other plants for the edges*
- *Containerized marginal and submerged plants*
- *Grit*

wood laid across its sides and a spirit level to check that it is level, lifting out the unit and rearranging the sand if necessary. Backfill with sand around the sides of the unit.

4 Pour water into the unit to a depth of 7.5–10cm (3–4in) to help keep it stable while knocking the stakes into the ground at 1m (3ft) intervals around the perimeter to match the height of the unit.

5 Place the log roll in position around the outside of the stakes. If slightly higher than the unit sides, make a slit trench in which to sink the base of the log roll so that there is a small gap left between the top of the unit and the top of the log roll.

6 Constantly check the unit to ensure that it remains both level and steady.

7 Cut the log roll to the length of the perimeter and use the galvanized nails to attach it to the supporting posts at the bottom and the top.

8 Fill the gap between the roll and the unit with sifted topsoil and tamp with the flat-ended length of wood. Firm with enough topsoil in the gap between the unit and the log roll to create a level planting area.

9 Plant alpines and other suitable plants in the narrow space between the log roll and the unit. After planting and watering the plants in, bale out the muddy water from the unit. Fill the pond to its full depth and plant with containerized marginals and submerged plants. Top-dress the plants around the pond edges with grit to conserve moisture and give a more finished look.

Right: Marsh marigold (*Caltha palustris*) makes a good choice for the edges of clay ponds as the roots are not invasive enough to penetrate the clay lining.

clay-lined ponds

Dew ponds seen in rural areas and on farmland probably started life as puddled clay ponds. In the days before modern lining materials, local clay and willing hands were all that was required to make a successful pond. Over the years, vegetation encroached from the sides, trees grew nearby and cattle tramped the edges when drinking from the pond. Consequently, these ponds shrank to a fraction of their original size and many disappeared altogether.

Clay soils are easy to identify. They hold water in the winter, they are heavy to turn over with a spade and tend to crack in summer. They are likely to have a high water table (see page 20), and it may be possible to excavate natural ponds on heavy clay soils to make an informal or wildlife pond, provided the lessons of the silted-up farm ponds are heeded.

Clay-lined ponds have a place where there is either heavy clay already on site or it can be bought cheaply from a local source. There is little justification for this method of construction for small garden ponds if the clay has to be imported some distance. It is never 100 per cent waterproof and the very slow absorption of water through its sides makes it more appropriate for large ponds.

tips

- *Avoid a site that is close to large trees, such as poplar* (Populus) *or willow* (Salix), *whose roots extend well beyond the outline of its branches.*
- *It is best not to lay a clay lining on chalky soil, since the chemical action of the chalk may increase the permeability of the clay.*

CHECKING SOIL FOR CLAY

Clay occurs in various colours, consistencies and depths. Few garden soils are composed of clay alone, and you will need to check whether the soil contains enough clay to make a puddled pond. The easiest way to do this is to roll a sample between the palms of your hands. If it falls to pieces it is unsuitable; it should stay tacky when moist and stain the palms as it rolls. Roll the sample into a ball and let it dry out to see if it has a tendency to crack. A more accurate test is to place a sifted sample in a glass of water with a teaspoon of salt. After shaking it up vigorously and leaving it for a day or two, it will settle into bands of sand, silt and clay from the bottom upwards. At least two-thirds of the sample should be comprised of the top layer of clay.

OTHER SOURCES OF CLAY

If clay is quarried locally for brickmaking, unfired bricks may be obtainable or there may be surplus loose clay that can be delivered inexpensively. Calculate the area of the pond

bottom and sides, then ask the supplier to estimate the quantity required on the basis that you will need to apply a layer of puddled clay at least 15–25cm (6–10in) thick.

If you are seriously considering using imported clay, sodium bentonite clay is likely to produce a better result and is much easier to install. Sodium bentonite is a patented clay product that swells to a volume of 10–15 times its dry bulk when wet. It is sold either loose in 50kg (110lb) bags or sandwiched in a geotextile mat in rolls 3.6m (12ft) wide. The mat is much easier to use than the loose product and is simply rolled out over the excavation in much the same way as a flexible liner.

INSTALLING A PUDDLING CLAY POND

1 Follow the excavation procedure for a pond with a flexible liner with marginal shelves and a deep zone on page 27, but make the shelves shallow – only 15cm (6in) under the surface – and well rounded, with a gentle gradient beneath. As the layer of natural clay will be a minimum of 15–25cm (6–10in) thick and there will be a 30cm (12in) soil layer on top of the clay, the excavation will need to be correspondingly deeper than for a flexible liner.
2 Line the hole with an inexpensive polythene sheet as a precaution against rodents, moles or worms.
3 Lay the clay pieces or bricks evenly over the excavation surface and tamp with a heavy tamper such as a road tamper (a thick pole with a heavy metal plate on the bottom), constantly keeping the clay moist while you are working.
4 Once the covering is complete, apply a layer of clay mud made from wettened clay pieces or bricks and smooth it off with the back of a spade to create a shiny, even surface.
5 Cover as quickly as possible with a 30cm (12in) layer of topsoil before introducing the water.

USING SODIUM BENTONITE MATTING

1 Roll the matting over the excavation to the edges. Overlap any additional rolls by 15cm (6in) and sprinkle loose bentonite between and over the join to strengthen the seal.
2 Cover the matting with a 30cm (12in) layer of topsoil.
3 In order to reduce the disturbance of this loose soil, add the water by resting the end of the hosepipe on a tile or slab and letting it in slowly. It will take several weeks before the fine particles of debris, silt and clay finally settle to the bottom and the water clears. It is important not to leave any clay exposed around the sides above the water line where it will dry out, crack and then erode back into the pond.

PLANTING CLAY-LINED PONDS

Clay-lined ponds are particularly vulnerable to root penetration by vigorous marginals plants, such as *Phragmites*, *Sparganium* and reedmaces (*Typha*). Choose less vigorous marginals such as marsh marigolds (*Caltha palustris*), Speedwell (*Veronica*) and water forget-me-nots (*Myosotis scorpioides*).

3 MOVING WATER FEATURES

Once you have introduced water into the garden, the urge to harness and exploit its full potential in terms of movement and sound will prove irresistible. Moving water and its associated sounds can have a significant effect on the character and atmosphere of a garden, and its nature needs to be in sympathy with the style of its setting. The sparkle, ostentatious yet controlled movement and lively sound of a fountain perfectly complement the clean lines and composure of a formal pool, in the same way as the soft ripples of a gently meandering stream over boulders and rocks subtly vitalize an informal pond.

planning requirements

The application of moving water in its various forms will be all the more successful if you spend time in advance observing all kinds of fountains and natural watercourses to study their effects and see how each type works best.

Constructing these features can be a highly creative process, involving relating the angle and direction of light to a fountain's droplet size and height, or weaving a stream through the garden to harmonize totally with its natural surroundings.

Unless the garden has a natural stream, any movement is dependent on the installation of an electrical pump to recirculate the water and these vary in their efficiency. Look carefully at the power consumption (wattage) on the box, as two pumps with the same output can have two totally different wattage ratings.

A tailor-made pump is available for almost every conceivable application. Installing a pump with an integral fountain head can be completed in a matter of minutes with the minimum of skill, but the services of an electrician will generally be needed to extend the power supply to a suitable point near the pond. A pump may only be the starting point – professional advice from a qualified electrical contractor will be invaluable in providing the right cables and sockets for other electrical items you may need, including lighting, filtration and a pond heater.

A feature incorporating moving water is not necessarily dependent on the presence of a mains water supply or steeply sloping ground, and if a stream, fountain or waterfall is well designed and constructed it will not place heavy demands on the domestic water supply in the summer.

Constructing a stream with rigid units

A simple stream can be made on sloping using preformed units for the course of the stream and for the header pool.

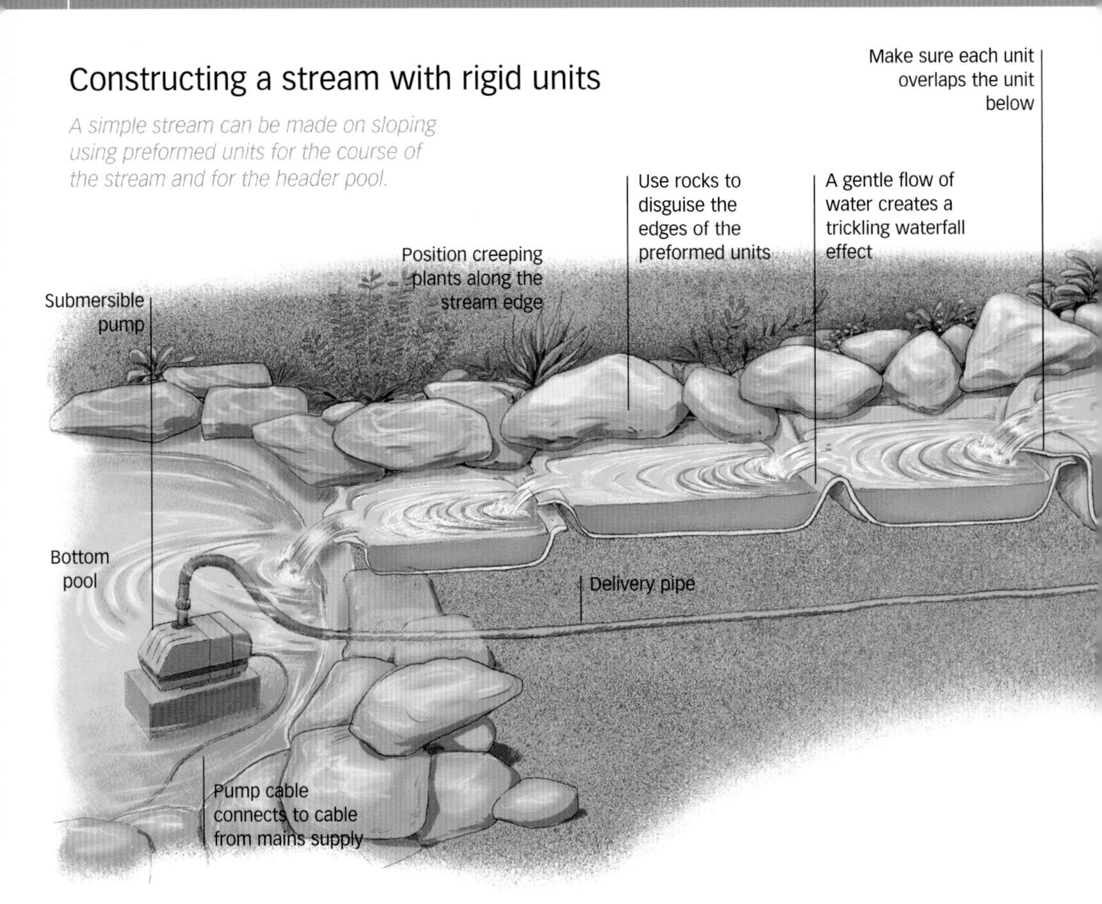

rigid stream units

Small streams and waterfalls are ideal candidates for using preformed fibreglass stream units and header pools. Larger and more ambitious watercourses using flexible liners and rocks to create the waterfalls require considerably more skill, since mortar is required to make the waterfalls watertight.

As with preformed pools, the main disadvantage with rigid stream units is in limiting the size of the stream. Some of the colours used are also extremely difficult to disguise and make look natural. But on the plus side, installation is relatively quick and easy.

The best rigid stream units are the fibreglass variety which are strong, have a long life-expectancy and are resistant to ultraviolet light. Stream units and rock pools are also made in PVC and vacuum-formed plastic moulds, which although cheaper have a limited life-expectancy and are much more easily damaged. All rigid stream units can be obtained in various finishes, such as pebbledash, textured rock and grit. They are best used amongst rocks on a sloping site where they are more easily blended into the scheme. If using several stream units, which are designed to overlap one another to form a longer stream, vary their direction for a natural look, since water seldom travels in a straight line down a slope.

Unlike with natural streams, the soil immediately surrounding moulded stream units is dry and therefore not suitable for marginal and moisture-loving plants that thrive in wet soil. There are, however, a number of creeping alpine plants that will grow in these dry conditions and help to disguise the artificial edges of the stream units.

Preformed rock pools are excellent for creating small pools at the top of the watercourse, known as header pools, which make a natural point of origin for a small stream. These rock pools contain a small reservoir of water even when the pump is not working and prevent a surge of water directly into the stream when the flow is turned on.

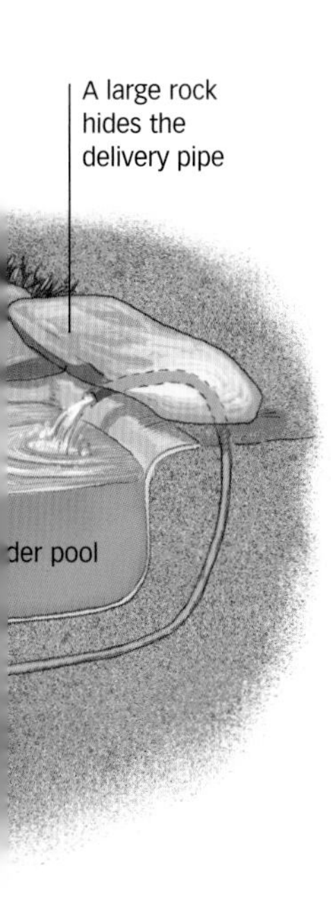

Right: Preformed stream units with a sandstone finish can be incorporated into a natural rock watercourse and, with a little ingenuity, be used to make effective waterfalls.

EXCAVATION AND INSTALLATION

1 Having measured the stream's length and purchased the necessary number of stream units and a header pool, mark out the route for the stream on the sloping soil next to the pond. Dig a level trench approximately 15cm (6in) deep and the same width and length as the first or bottom stream unit. Line the trench with a 5–8cm (2–3in) layer of soft sand. Position the unit firmly and level onto the sand, with the outlet projecting over the bottom pool by 8–10cm (3–4in) and enough water added to keep the unit stable. Bury the delivery pipe from the pump 15cm (6in) deep alongside the stream.

2 Install as many intermediate sections as required in the same way until the top of the stream is reached. Ensure that the outlet for each of the units overlaps the unit below.

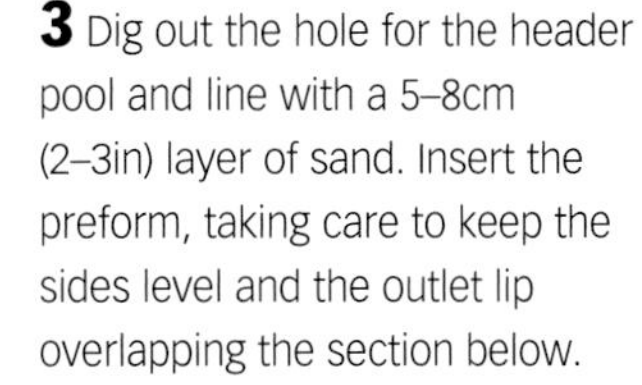

3 Dig out the hole for the header pool and line with a 5–8cm (2–3in) layer of sand. Insert the preform, taking care to keep the sides level and the outlet lip overlapping the section below.

4 Use a backfilling of sand or sifted soil tamped around them to secure the units in place. Run the delivery pipe from the pump into the header pool and disguise it with rocks. Keep the end of the pipe above the water line of the header pool to prevent it siphoning back the water to the bottom pool when the pump is turned off.

5 Make the connection of the integral cable from the submersible pump to the cable from the mains supply with a waterproof connector. Ensure that the mains supply cable to the pond is protected by an RCD device (contact circuit breaker). Turn the pump on to check that the stream is flowing satisfactorily and then make any final adjustments to the levels of the units.

6 Position a few strategic rocks throughout the watercourse and plant creeping plants along the stream edge.

CONSTRUCTING A STREAM WITH RIGID UNITS

You will need

- *Tape measure*
- *Rigid stream units and header pool*
- *Wooden pegs and string for marking out*
- *Spade*
- *Trowel*
- *Wheelbarrow*
- *Soft sand*
- *Spirit level*
- *Length of plastic corrugated or reinforced flexible pipe or hosepipe 2–2.5cm (¾–1in) in diameter to act as delivery pipe from the pump (use a 2.5cm (1in) pipe if the distance to the top of the stream is more than 3m (10ft))*
- *Medium- to small-sized rocks*
- *Submersible pump*
- *Approved waterproof connector*
- *Contact circuit breaker*
- *Selection of creeping plants*

streams with a flexible liner

This type of stream construction allows for more variations in design. Depending on the change of height and length of the proposed stream, it may be constructed with one piece of liner or several overlapping pieces. In the case illustrated, the stream is being constructed on a relatively flat site using a single length of liner.

EXCAVATION AND INSTALLATION

1 Carefully select the site for the stream. In this case, the stream will lead the eye out into the surrounding countryside in the distance.

2 Use pegs to mark out the route of the stream. Use a spirit level on top of a straight-edged length of wood to check that they are level. In this instance, the stream will take the form of a long, narrow pool, which is level along its length so that the water will remain in the stream when the pump is off. As long as the outlet into the base pool is lower than the edges of the stream, the stream will flow when the pump is turned on. The outlet into the base pool is the point to which all other levels must relate, known as the datum level.

3 Remove the turf to the width of the proposed stream. Keep the level pegs in place at this stage so that you can check whether any building up of the stream edges will be necessary.

4 Remove the soil in the centre of the stream to a depth of 30cm (12in), and temporarily store on a polythene sheet nearby for later use. Make shelves 15cm (6in) deep and 7.5–15cm (3–6in) wide along the sides, or sections of the sides, to make it easier to position any rocks or to create planting pockets. Remove any sharp edges or stones and rake the surface of the excavation until it is smooth.

5 Drape the underlay along the length of the stream. Use small rocks to secure it in place until the liner is inserted.

6 Drape the liner in position and temporarily hold in place with rocks that will be used later along the stream edge. Remove any stones or sharp edges, then replace the soil dug out of the stream to a depth of 7.5–10cm (3–4in). This will prevent light from deteriorating any of the more inexpensive liners, although butyl is resistant to ultraviolet light (see page 27). Position plants at suitable points in the stream, which over time will have a filtering effect on the base pool. Use rounded pebbles or beach cobbles to top-dress the soil and prevent erosion taking place when the water movement starts.

7 Before finalizing the positioning of the rocks, bury the length of corrugated plastic pipe alongside the stream to feed the pumped water to a rocky outcrop, which will act as the source of the stream. Within 4–5 months the rocks will have matured and become a home to moss and plants. The planting in the stream itself will quickly soften the stream edge.

8 This stream was constructed in the early autumn, and by the following late spring it is already looking mature with the flowers of marsh marigolds (*Caltha palustris*), buds of the Siberian iris (*Iris sibirica*) and the creamy, striped leaves of the yellow or flag iris (*Iris pseudacorus*). To add further architectural impact, the slow-growing fan palm (*Trachycarpus fortunei*) has been planted in a prominent position by the patio window.

9 Looking back to the source of the stream, the newly enlarged base pool has taken on a lush feel with its surroundings of New Zealand flax (*Phormium*) and bamboo. All that remains is to finish making the connection from the stream into the pool below with some pieces of flat rock.

CONSTRUCTING A STREAM WITH A FLEXIBLE LINER

You will need

- *Wooden pegs and string for marking out*
- *Spirit level*
- *Straight-edged length of wood*
- *Spade*
- *Wheelbarrow*
- *Large polythene sheet*
- *Rake*
- *Spun geotextile membrane underlay*
- *Small rocks*
- *Flexible liner*
- *Selection of oxygenating and marginal plants*
- *Trowel*
- *Rounded pebbles or beach cobbles*
- *Corrugated flexible plastic pipe 2.5cm (1in) in diameter*
- *Submersible pump*
- *Approved waterproof connector*
- *Contact circuit breaker*
- *Pieces of flat rock*

Top: Dig the marginal shelves to a depth of 22cm (9in) and the centre to 37cm (15in). *Above*: Drape fabric underlay into the stream to protect the liner from sharp stones.
Centre: Position the liner and hold it in place with flat rocks along the edge until you are satisfied. Then you can add the soil and water and finalize the positions of the rocks. *Bottom:* Less than a year after its installation, the planting is in flower and the stream looks mature. The flat-topped rocks act as stepping stones across and alongside the shallow water.

Left: A ring fountain works perfectly in this simple, circular pool. The visual balance between the jets is vital so they should be cleaned meticulously.

fountains

Fountains are at their best in high summer when they bring refreshment to hot, dusty atmospheres. In addition to their ornamental value, they are extremely beneficial in fish ponds, where they provide valuable oxygen if left on during hot, sultry nights when the surface water cannot absorb enough oxygen on its own. It is worthwhile researching the many different styles of fountain – particularly their height and spray pattern – that can be fitted into a pool, depending on its size and design, since the standard plastic fountain nozzle generally supplied with pump kits is really rather unimaginative.

HOW THEY WORK

The simplest fountain operates directly above a submersible pump, which sits on a shallow plinth on the pool bottom. A simple fountain kit is supplied with a submersible pump and consists of a small length of rigid fountain extension pipe which has a flow adjuster; a T-piece extension to permit a second pipe to a waterfall or filter; and a push-on plastic spray nozzle, which fits on top of the extension pipe. The length of rigid pipe supplied with the fountain nozzle is designed for a pond 46cm (18in) deep. If the pond is much deeper, it will be necessary to elevate the pump so that the fountain nozzle clears the surface. If the pond is much shallower, the pipe can be shortened with a hacksaw. A remote-controlled pump is now available which can be operated from the house if the fountain needs to be turned on and off regularly.

The height and width of spray can be regulated by adjusting the flow adjuster. On exposed sites, keep the height of the fountain spray no higher than the radius of the base pool.

MAINTENANCE

In small, shallow ponds it is relatively easy to reach the adjuster and to clean the spray nozzle and pump strainer, which clog up frequently. For larger ponds it is better to site the pump close to the pool side and connect the fountain to the pump with a length of flexible pipe which lies on the pool

- *Fountains are extremely effective when the water droplets are caught in sunshine or artificial light. Site the fountain with a dark background and preferably where the sun will catch the spray.*

Right: Tiered ornamental bowl fountains make a bold contribution to a formal garden, remaining a focal point even when the fountain is turned off.

bottom. Choose a dark-coloured pipe to make it less conspicuous.

Algae, particularly the filamentous types that make up blanketweed, is the bugbear of small fountain pools, clogging up the strainer and fountain jets. Remove these to clean frequently in order to maintain the fountain height, cleaning the jets with diluted vinegar to reduce the lime scale at the same time.

TYPES OF FOUNTAIN JET

Single spray or plume spray

This is one of the simplest types, ideal for small pools, and eliminates the need for cleaning fine nozzle jets. The narrower the diameter of the fountain pipe, the higher the spout will lift. A geyser-like effect can be achieved by keeping the top of the outlet pipe just below the water surface.

More expensive geyser jets are available which give a very frothy plume. These are relatively expensive and require a more powerful pump.

Bell jets

These produce a thin hemispherical film of water that is only suitable for very sheltered ponds, since it is broken up in any breeze. They make good features for indoor or conservatory fountains.

Two-tiered sprays

By altering the arrangement and size of the nozzle holes of a simple rose spray, it is possible to produce tiers of spray which are very effective in raised pools or formal schemes, where the sophisticated patterns are set off by simple surroundings.

Surface ring jet

This fountain may have a variety of spray patterns emanating from the water surface without any stem being visible. Any variance in the water level could spoil the effect by submerging the jets or exposing the fountain stem.

Whirling spray

A horizontal whirling wheel is driven round by the pumped water in much the same way as some lawn sprinklers. The greater the water pressure, the faster the arms rotate.

Left: A stone trough and mask make a simple wall fountain. The reservoir, pump and pipes are hidden under the trough.

wall fountains

Small enclosed town gardens, courtyards, patios and conservatories make ideal situations for wall fountains where there is often insufficient room for a free-standing pond or fountain pool. The various forms of fountain outlet, mainly masks and gargoyles, lend themselves to creating small architectural features. These should be chosen with great care, since they tend to dominate, even when the fountain is turned off. A wall fountain can make a strong focal point at the end of a path, particularly if it is caught by sunlight and can be seen from the house.

DESIGN FEATURES

As with all fountains, a reservoir is needed to act as a sump for the pump. This may take the form of a wall-mounted container a short distance beneath the spout or a pool at the base of the wall. Generally it is better to have a base pool with ample reserves of water to combat evaporation loss and splashing. A wall-mounted reservoir capable of holding adequate water reserves is likely to be heavy and require strong supports.

The design of the base pool allows for variation. A raised pool with a wide coping offers a place to sit and enjoy the sound and feel of the water. If the walls of the raised pool are made of the same material as the wall on which the spout is mounted, it will appear as if it has always been part of the wall design.

Self-contained wall fountains are sold complete with a very small pump and require no more than screwing to the wall and the pump cable connecting to a socket. They are generally small features and could look out of scale on a large expanse of wall.

TECHNICAL CONSIDERATIONS

One of the main technical problems to be overcome with larger wall fountains is the disguising of the delivery pipe from the pump to the wall spout. If the wall has an adequate cavity, the hose can be pulled through after making the entry and exit holes. If access is possible behind the wall, the method below can be used. Where neither of these options is possible, the pipe will need either to be disguised with wall trellis and climbing plants or hidden behind a terracotta tile drain cut in half and mortared to the wall, in a vertical line from below the spout to the base pool. The plumbing can be hidden by building a false wall with a small gap in front of the existing wall, at the same time providing a more secure construction to support any heavy wall basin.

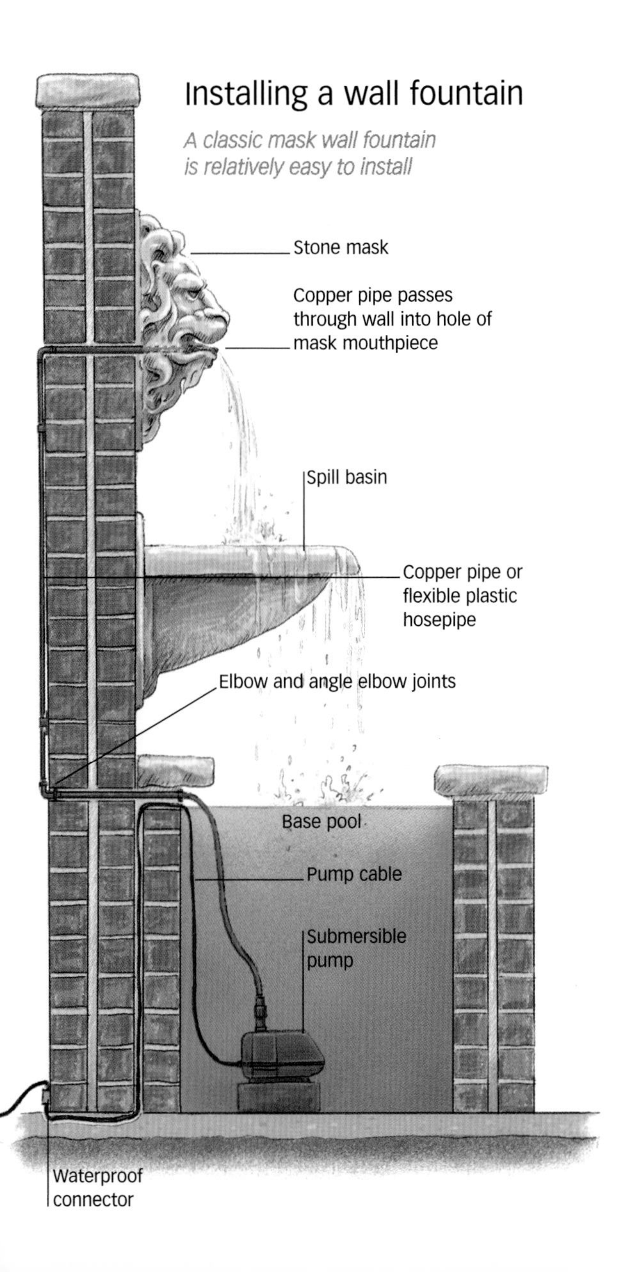

Installing a wall fountain

A classic mask wall fountain is relatively easy to install

INSTALLING A WALL MASK FOUNTAIN

You will need

- *Reconstituted stone mask*
- *Base pool*
- *Electric drill and a 1.5cm (½in) masonry bit*
- *Copper pipe 1.5cm (½in) in diameter*
- *Hacksaw for cutting copper pipe*
- *Small bag of ready-mix mortar*
- *Mortaring trowel*
- *Bucket for mixing mortar*
- *Elbow joints to fit copper pipe*
- *Screwdriver*
- *Flexible plastic hosepipe*
- *4 x 1.5cm (½in) hose clips or 2 angle elbow joints to fit copper pipe and elbow joints*
- *Submersible pump with flow adjuster*
- *Approved waterproof connector*

INSTALLING A WALL FOUNTAIN MASK

1 Mark the position of the mask mouthpiece on the wall and drill a hole through the wall. In order to hide the delivery pipe behind the wall, drill a second hole in the wall just above the water line in the base pool.

2 Cut a piece of copper pipe long enough to pass through the wall into the hole of the mouthpiece. Put the pipe in place to provide temporary support for the mask, leaving at least 2.5cm (1in) projecting out from either side of the wall.

Above: Creating an element of surprise in this tiny niche, the simplest of wall fountains has been created in this dry wall. The tiny pump is hidden by cobbles in the container.

Below: It is important to choose the right-sized pump. This wall fountain needs a powerful pump to achieve this effect.

Right: Matching the existing brickwork exactly means that this wall fountain sits harmoniously within its surroundings.

3 Spread a thin layer of mortar over the back of the mask, after first moistening the surface for better adhesion. Make sure that the mortar is not too sloppy or it will ooze out of the sides.

4 Present the freshly mortared mask to the wall, sliding the mouthpiece over the copper pipe for support until the mortar dries.

Press firmly and remove any surplus mortar. Remove any spilt or conspicuous mortar with a damp sponge before it hardens. Once the mortar has hardened, make any adjustments to the length of the copper pipe to create the best spout.

5 Attach the end of the copper pipe to an elbow joint. If using flexible hosepipe, attach the elbow joint to a vertical length of hosepipe, then attach the other end to a second elbow joint, using the hoseclips to tighten the hosepipe onto the elbow joints. Join the bottom elbow joint onto a short length of copper pipe that feeds back through the lower hole in the wall and connects with the submersible pump in the base pool. If you are using copper pipe, use angle elbow joints to fix the copper pipe to the elbow joints. Secure any pipework by screwing brackets to the back of the wall.

6 Connect the submersible pump to the mains electric supply with the waterproof connector, then test the system for the rate of flow. Adjust the flow adjuster if necessary.

Left: Lines of cobbles mortared into the base of this formal canal complement the more loosely arranged shingle and gravel in the adjacent paths and beds.

canals

Canals are perfect for linking water features in formal gardens such as raised pools or fountains with shallow refreshing movement, and for drawing the eye to sculptures, urns and other decorative features. Keep the canal width to the scale of the garden, erring on the narrow side rather than the wide. A canal 30cm (12in) wide will have quite a dramatic impact, particularly if accentuated by columnar plants or clipped hedges and topiary.

Canals and canal pools should harmonize with the surrounding walls and paving, providing a perfect setting for restrained planting and containers. Explore the range of suitable materials available. Concrete is an ideal choice for construction, given the shallowness and rigid definition of a canal, while glazed tiles used for edging create a wonderfully cooling effect.

This type of feature requires a great deal of care in its construction with close attention paid to checking levels, depths and proportions.

tip

- *Reinforcing fibres prevent the concrete used in the canal from cracking if there is any future ground movement. The more traditional method of incorporating galvanized mesh into the concrete base will also help prevent large cracks but it is not such a reliable method.*

EXCAVATION AND INSTALLATION

1 Mark out the width and length of the proposed canal on a level site. Excavate the area to a depth of 15cm (6in) for the canal and a further 5cm (2in) for the base layer of soft sand. Add the sand, then drape the polythene sheet across the excavation to provide a protective base on which to bed the concrete.

2 Check that the sides of the excavation are level using the straight-edged length of wood and spirit level. Mix up the concrete using 2 parts sharp sand and 1 part cement, adding reinforcing fibres during the mixing following the manufacturer's instructions. Pour onto the polythene sheet and level by tamping with the flat-ended length of wood.

3 After the concrete mix has hardened in a day or two, construct the side walls using common bricks mortared onto the base. If brick pavers are to form the surrounding edge, two courses of bricks will provide an adequate depth. Ensure that the brick courses are dead level as construction proceeds.

4 Once the mortar has hardened, lay the coping bricks at right angles on top of the side walls and mortar into place, allowing an overlap over the water of approximately 5cm (2in). After allowing the mortar to harden, skim the inside of the canal with a cement mix containing reinforcing fibres, using a plasterer's float. The thickness of the skim should be 5mm (1/5in) on the side walls, with two layers each 5mm (1/5in) thick applied to the base.

USING A PAVING SLAB BASE

An alternative method of constructing a canal is to use a paving slab base instead of concrete, which requires fewer building skills. For short canals, a paving slab 90 x 60 x 5cm (3 x 2ft x 2in) would be adequate placed singly or side by side along the length of the canal, provided there was a sufficient skimming of cement between the joints and over the total surface area of the slabs and walls. Concrete walling blocks 46 x 23 x 7.5cm (18 x 9 x 3in) could be used to form the canal sides instead of bricks and a decorative paving stone 5cm (2in) thick used for the edge. If employing this method for long stretches of canal, the skimming should contain reinforcing fibres in order to strengthen the seal between the joins in the paving.

INSTALLING A CANAL

You will need

- *Bamboo canes and string for marking out*
- *Spade*
- *Wheelbarrow*
- *Soft sand*
- *Heavy-duty polythene sheet*
- *Straight-edged length of wood*
- *Spirit level*
- *Sharp sand*
- *Cement*
- *Reinforcing fibres*
- *Flat-ended length of wood, 5 x 5 x 60cm (2 x 2 x 24in), for tamping*
- *Common bricks*
- *Bricklayer's trowel*
- *Ready-mix mortar*
- *Coping bricks (engineering bricks or brick pavers)*
- *Plasterer's steel float*

Formal canal

This cross-section shows the construction materials needed to build a canal or rill.

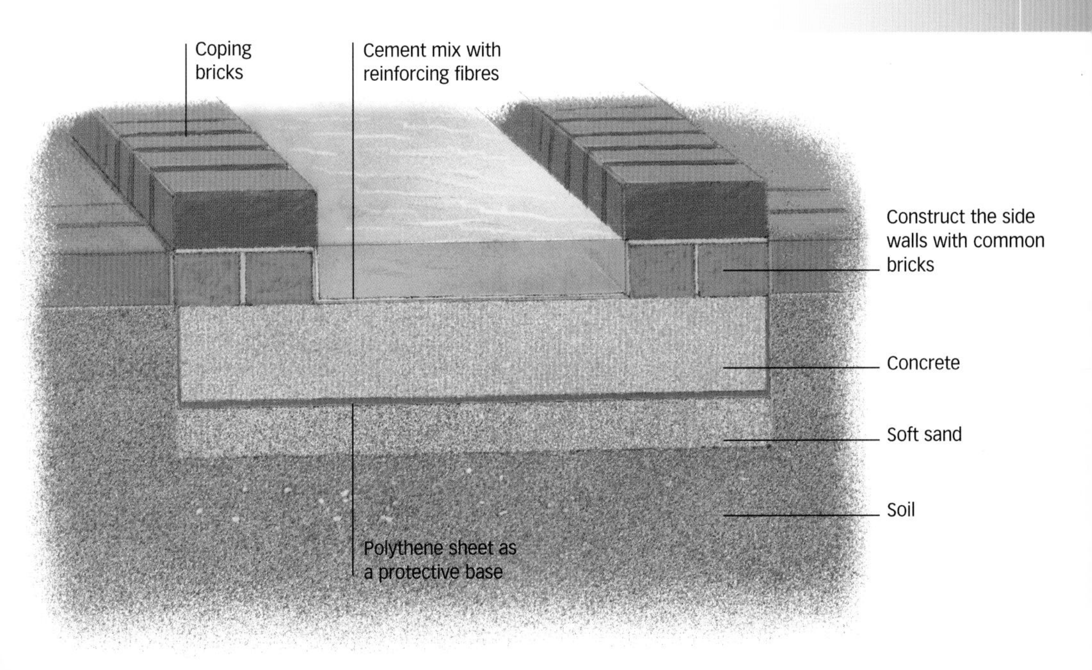

Left: This hexagonal planter has been turned into a cobble fountain by sealing the inside and placing a mesh across the top to support the cobbles. The pump is placed under the mesh and the cable hidden in the foliage. *Right*: This brimming urn is supported on brick piers over a sunken reservoir that holds the pump. The water is fed through a hole in the base of the urn.

reservoir features

The variety of efficient submersible pumps and the increasing range of larger containers have led to more and more innovative ways of creating moving water features for the smallest space in the garden or conservatory. The basic requirements are the proximity of an electricity supply and a reservoir large enough to cope with the recirculation of water where there could be evaporation loss. Once installed and filled, these features place little demand on the mains water supply other than the occasional topping-up. They also require little maintenance, unlike pools which are dependent on plants or filtration to maintain clear water, and are surprisingly inexpensive to run.

The cobblestone fountain is a simple feature appropriate for both formal and informal settings, and can be as small or as large as the garden dictates. Its effect can be varied considerably simply by changing the arrangement of cobbles at the outlet or by raising or dropping the level of water in the reservoir, which influences the sound of the water falling in from the cobbles.

INSTALLING A COBBLESTONE FOUNTAIN

1 Choose a site that can be easily seen from a favourite viewing window and clear an area of ground approximately 1–1.2m (3–4ft) square. Measure the diameter and depth of the tub or bin and dig a hole slightly larger than these measurements.

2 Lower the reservoir into the hole so that it is just below ground level. Check that the rim is level with a spirit level on a straight-edged length of wood. Firm the surrounding soil by tamping it into the gap between the reservoir and the perimeter of the hole. Once the reservoir is secured, create a saucer-like depression 7.5–10cm (3–4in) deep extending to a radius of 1m (3–4ft) in the surrounding ground around the reservoir rim.

3 Drape the polythene sheet over the prepared area and temporarily secure with cobbles. Cut out a hole over the top of the reservoir slightly smaller in diameter than the reservoir rim.

4 Place the paving stone in the bottom of the reservoir to prevent any debris from clogging the intake strainer on the pump and lower the submersible pump onto it. Connect the length of pipe to the pump outlet.

5 Place the galvanized metal or wire mesh over the reservoir. Cut a small hole in the centre of the mesh and thread through the outlet pipe from the pump.

6 Lodge the pipe between cobbles, then arrange more cobbles around it on the wire mesh. If the pipe is flexible, ensure that the arrangement of cobbles maintains the pipe in an exact upright position in order to achieve the effect of the water falling back on itself. If a rigid pipe is used, the cobbles play less of a role in keeping the spout of water upright. Position just enough cobbles at this stage to assess the effect when the pump is turned on, since any adjustment required to the flow adjuster on the pump will involve removing the cobbles and mesh. Partially fill the reservoir and turn the pump on to check the effect.

7 Once you are satisfied with the rate of flow, fill up the reservoir. If the outlet pipe is too conspicuous, trim it to the required length, then finalize the arrangement of cobbles depending on the effect you want. Mounding the cobbles over the outlet dissipates the water into several small sprays, while leaving it clear results in a single plum of water. Finally, arrange the remaining cobbles over the surrounding mesh.

Installing a cobblestone fountain

Cobble fountains are ideal where there isn't enough space for a full-size pond

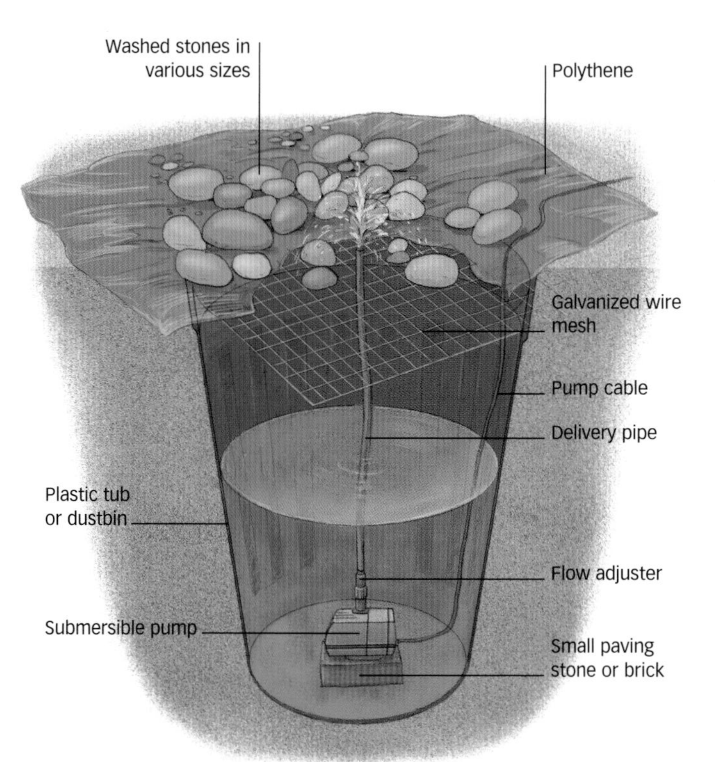

INSTALLING A RESERVOIR FEATURE

You will need

- *Spade, rake and trowel*
- *Wheelbarrow*
- *Large plastic tub or bin*
- *Spirit level*
- *Straight-edged length of wood*
- *Flat-ended length of wood, 5 x 5 x 60cm (2 x 2 x 24in) for tamping*
- *Polythene sheet*
- *Various washed cobbles*
- *Small paving stone approximately 5–7.5cm (2–3in) deep to fit within the base of the reservoir*
- *Submersible pump with flow adjuster*
- *Rigid or flexible pipe to fit the pump outlet and long enough to clear the top of the reservoir (a smaller-diameter pipe forces the water higher; a larger diameter makes it lower)*
- *Galvanized metal or wire mesh 10cm (4in) wider than the diameter of the reservoir top*
- *Wire cutters*
- *Approved waterproof connector*

4 DECORATIVE FEATURES

There are many features that can be added to a pond at a later stage to enhance its appearance. The type of edging you choose has the greatest impact on the overall design of the pool, whether formal or informal. For the latter, the most ambitious to construct is a decking edge and this is much more in keeping with a natural, informal setting than a paved edge. Other features offer improved access as well as aesthetic pleasure in the form of stepping stones or bridges, or lighting that can extend the enjoyment of the pond into dark, winter evenings.

edging

Paving stones are the traditional choice for surrounding small ponds, but there are now many other suitable materials that offer greater creative scope. These materials demand sound construction techniques and a great degree of accuracy, particularly with regard to levels. The strength and stability of materials is also an important issue, since areas around a pond subject to heavy usage will quickly wear or subside if the foundations are inadequate.

FORMAL EDGING

Formal pools are best suited to paved edges, which give a crisp definition to the pond. There is a vast range of paving available both in pre-cast concrete slabs and natural stone. While the latter blends well with mature settings, it can be a potential hazard near water where it quickly becomes covered with algae, making the surface very slippery. Pre-cast slabs on the other hand are available with non-slip surfaces and are less prone to algae growth.

Where formal designs require tightly curving or circular outlines, bricks or brick pavers make a good outline, especially when the colour contrasts with any other paving there may be around the pool.

Natural stone

Reclaimed natural stone slabs are the most expensive paving slabs. Ensure that they are of an even thickness or they will look odd when overlapping the water surface. They are generally supplied in rectangular sizes of mixed dimensions rather than all the same size.

Brick

If kept above the water line, facing bricks can be used for paving where there is no heavy wear. For heavy wear areas or where there is contact with water, use engineering bricks or brick pavers which are much harder than facing bricks and also waterproof.

Textured concrete slabs

Available in several sizes and colours, these can be laid with confidence on a level surface, since they all have exactly the same thickness.

Crazy paving

A common and relatively inexpensive natural-stone edging which is versatile enough to be used for both straight and curving outlines.

Setts

Setts form useful edges and are versatile in the same way as bricks for tight curving outlines. Setts are available in granite, sandstone and concrete in either 10cm (4in) or 15cm (6in) cubes. They must be embedded in mortar for rigidity.

INFORMAL EDGING

Natural edging in the form of plants or grass is obviously highly appropriate for informal ponds, but it presents problems of access and maintenance.

Grass

Grass edges should be prevented from crumbling by underpinning with a solid foundation (see pages 30–1) or combining with vertical timber rounds. A mixture of grass and other materials can look effective around large ponds, particularly where there is variation in the surrounding landscape and only one access point is required. Provided the grass cuttings are boxed off and are not allowed to blow into the water, grass makes a relaxing and natural edge. Keep a definite line by trimming it during the summer to prevent it from growing under the water.

Timber rounds

Hardwood or treated softwood rounds 9cm (3½in) in diameter can be used to form an edge between the water and grass to prevent the grass from eroding. The tops of the rounds can be cut off to just below grass level to allow for mowing, or left proud so that soil can be banked up at the pool side.

ROCK EDGING

Because of their natural association with water, rocks enhance the appearance of informal pools and streams when used as part of the edging. There are many types of rock available from major suppliers, but as a general rule, use local stone on the grounds of cost and their more natural effect. The two most common rocks are sandstone and limestone.

All sorts of materials can be used to edge water features. *Far left:* Bricks are useful for a small circular sunken pool whereas timber rounds (*Centre*) make an innovative edge to a raised pool. *Above:* For an informal and natural effect, both harder forms of sedimentary rocks make successful edges.

Sandstone

Sandstone has a pleasing and unobtrusive colour that ranges from a light yellowish grey to a dark reddish brown, depending on its origin. Both colours tend to change with time as the stone weathers. It is a strong stone but soft enough to weather into interesting stratified patterns and porous enough to absorb moisture. This porosity can be a disadvantage in areas of severe frost, causing the stone to flake or split. A complete edging of a soft sandstone around a small, lined pond would increase the drop in water through evaporation from its porous surfaces. Harder sandstones such as millstone grit have very attractive colourings, are less porous and are therefore ideal for edging water.

Limestone

This stone is usually grey in colour, and although porous, it weathers into attractive patterns of cracks and crevices. If partially submerged, a small amount of lime will leach out. This is only a problem for very small pools in a hard-water area that are regularly topped up, resulting in the water becoming very alkaline. Like the sandstones, there are many variations in the colour and character of limestone, but cream colours are more prone to splitting.

Quartzite

This is a harder, less porous rock and can be found in interesting colours folded into the rock, exploited to great effect by Japanese craftsmen in the Kyoto water garden in Holland Park, London.

Slate

For sheer drama, slate is particularly effective when used in watercourses, where its colours are highlighted in a glistening sheen and marbled with white veins.

Pebbles and cobbles

Pebbles or cobbles make one of the most successful edges for informal water features and are available in a wide variety of sizes. They come in a range of appealing colours including white marble and honey quartzite. The best type for ponds are beach cobbles, which are available commercially. In many areas, it is illegal to collect cobbles from the local beach, and only certain beaches are licensed to remove cobbles for decorative use.

Making a cobble beach

A sloping pebble beach is attractive and allows wildlife easy access to a pond.

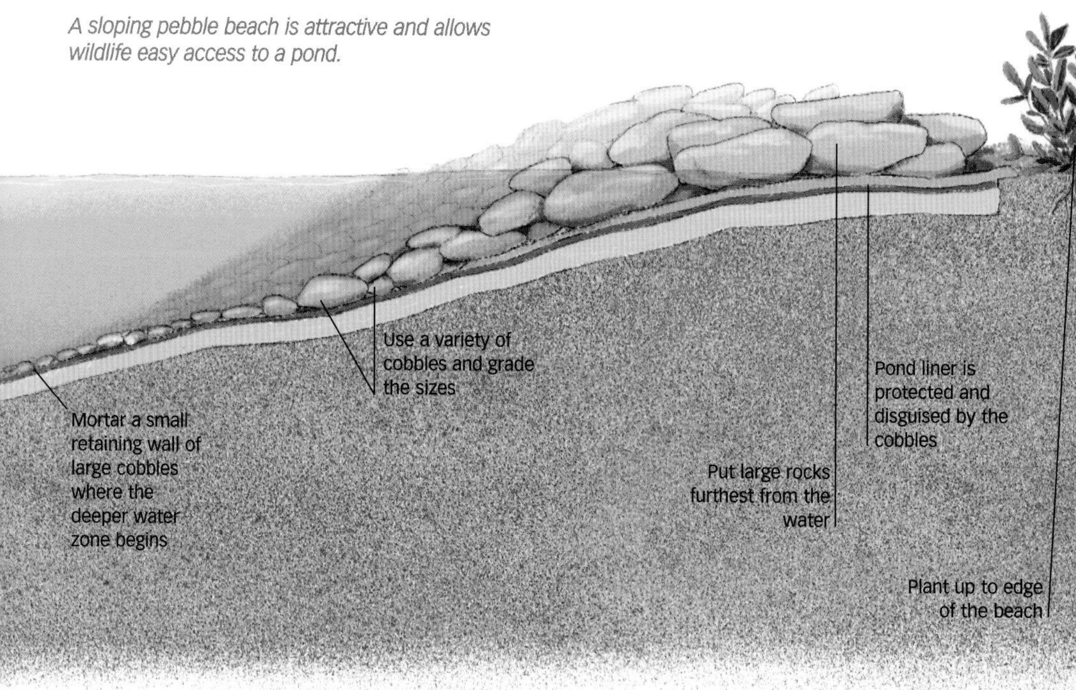

Loose chippings

Angular aggregates and crushed gravels are often too small for using on their own as an edging because they can be easily carried on the feet when damp. If angular rocks are used as an edging, loose chippings make a useful gritty surface between the rocks. They should only be used above the water line and not allowed near flexible liners because of their sharp edges. Since they are easily lost in the soil, they are best spread onto sheets of permeable membrane sold for landscaping which suppress the weeds but allow water through.

Decking

One of the great advantages of using wooden decking as an edging is that you can project it out over the water like a jetty, helping to create the illusion of running water and forming a substantial sitting area. The pool edge is hidden from sight underneath the decking. Decking is ideal for dry situations, such as the sunny side of the house where the algae are baked off the wooden surface in prolonged sunshine. Long-lasting hardwood should be used whenever possible, but if the cheaper softwoods are used make sure the wood has been adequately treated with a preservative beforehand. The decking planks and tiles are better if they are ribbed with grooved surfaces to prevent slipping. Tiles with diagonally arranged timbers make an attractive pattern on large surface areas.

Pebble beaches

By making a portion of the edge of a pool a beach or shallow gradient, wildlife are ensured access to the pond throughout the various fluctuations of water level in summer and winter. Pebbles or round cobbles make an ideal covering for shallow beaches, preventing the edges from turning to mud and the water from becoming cloudy through the activity of birds.

Use a mixture of different sizes of cobble to make a more interesting and natural-looking beach, and grade the sizes so that they increase in diameter from below the water line to the drier margins. In order to prevent them from rolling to the bottom of the pond, mortar a small retaining wall of larger cobbles at the point where the shallow edge finishes and the deeper zone begins.

Pebbles disguise liners very effectively and prevent any deterioration through the action of ultraviolet light. The liner can also be protected from puncturing by dogs' claws if the cobbles are secured underneath the water line with concrete.

Below: A shallow beach of pebbles is both child- and wildlife-friendly and here provides a smooth contrast to the texture of the plants.

Left: A timber bridge blends into an informal setting more easily if it does not end abruptly with vertical posts. Here the sloping handrails bring the structure gradually into the planting . *Right:* These stepping stones provide an easy means of access to the larger decking 'island'.

bridges

A good bridge serves a function as well as being ornamental, and should not look like an afterthought without a real purpose. Apart from providing a link to pathways, bridges offer excellent vantage points for viewing the water and appreciating new reflections. Arched bridges, however, can become obtrusive and unrestful, creating an unnecessary humped-back obstacle to an otherwise leisurely walk.

Wooden bridges built close to the water surface are the simplest and most effective, particularly in an informal garden. Planks arranged longitudinally over the water make the bridge look longer and more inviting; arranging them at right angles makes the bridge look shorter, signalling you to stop at the bridge. Only use handrails where they are necessary since the view is generally better without them. If the handrail is vital for safety, try to merge the ends of the rail into planting rather than having abrupt vertical ends.

Informal water gardens have a tendency to become dominated by lush foliage in the margins. To give increased dominance to the bridge, position it across a narrower section of water. The bridge should be built well into the surrounding banks, which has the effect of making a narrow piece of water appear wider.

CONSTRUCTION GUIDELINES

The maximum span for an unsupported wooden bridge is 2.4m (8ft). If it is longer than this, you will need to erect supporting piers in the water. The minimum width for a bridge is 60cm (2ft), and in a modern garden it should be wide enough to cater for maintenance equipment, such as a lawnmower. When placing a long single wooden joist over the water, ensure that the slight curve normally in evidence is uppermost, so that when screwing or bolting the cross pieces together the strain tends to straighten the bow in the wood with time. Cross timbers measuring 10 x 5cm (4 x 2in) placed every 90cm (3ft) between the joists help to prevent twisting or warping of the main weight-bearing joists.

Make level concrete foundations to support the bridge at each end using a mixture of 1 part cement to 4 parts 2cm (¾in) ballast. In order to secure the bridge to the foundations, bed bolts with their screw ends uppermost into the soft concrete before it hardens.

After the concrete has set firmly, the bridge joists are connected to the bolts by right-angle brackets. Non-slip or ridged planks 15cm (6in) wide are then cut into 60cm (2ft) lengths to form the floor, overlapping the supporting joists after being screwed down with countersunk 5cm (2in) screws.

stepping stones

Stepping stones offer an alternative method of crossing water, even if only to linger midway to enjoy the sensation. They make any expanse of water more interesting and lead the eye from one part of the garden to another, particularly if they are roughly in line with the main viewpoint rather than at right angles to it. Stepping stones should not be used in deep water or approached from steeply sloping ground. Their placing should ensure easy and gentle strides without having to watch your step too carefully when gazing into the water. If in doubt as to how far apart to put them, err on the side of caution and place them closer together. Once the crossing place is finalized, a sturdy foundation should be provided, preferably before the liner is inserted.

PAVING SLABS

These make good stepping stones, particularly reconstituted slabs with riven or non-slip surfaces. Although used mainly in formal garden schemes with square or rectangular outlines, manufactured concrete stepping stones are available in round or irregular shapes with riven surfaces for more informal schemes. Circular paving slabs make a dramatic impact in modern designs, echoing the roundness of water lily leaves. Whatever shape is used, ensure that the individual stones are wide enough to allow for comfortable standing room.

BRICKS

Bricks make good piers for supporting paving used as stepping stones. In pools no deeper than 60cm (24in), courses of four bricks butted at right angles to each other allow a paving stone of 46cm (18in) square to be placed on top with a slight overhang, without any bricks having to be cut. When the brick courses reach above the water line, the paving stone can be placed squarely onto mortar dabbed onto the top of the pier and checked for level with a spirit level as it is tapped into the soft mortar.

BOULDERS

Large round-topped or flattish natural boulders are ideal in informal water features such as streams and wildlife pools. Bog gardens also require an informal dry access that is not too artificial in an otherwise natural planting scheme. Choose non-porous boulders such as granite since softer, porous stones such as some freshly quarried sandstones may crumble in time with the action of frost and attract slippery algae. Rounded boulders look good when they are partially submerged and are most appropriate at narrow points in a stream, where subtle placing gives no hint of deliberate bridging and so looks natural. The tops of waterfalls also make excellent points to introduce a few stones in the clear, shallow water before the water tumbles over the fall.

Far left: A effective way of continuing the line of a path through a pool is to use a stepping stone of the same material in the water.
Left: Leave a shallow, flat ridge in the pond so that the brick piers for the stepping stones are no more than 45cm (18in) high.
Above: Flat-topped pieces of sandstone laid at random across a shallow stream make an informal crossing point.

Informal stones need to be bedded onto a mortar base to prevent any danger of wobbling. Ordinary soil, although initially appearing to help stabilize the stone, will become soft mud very quickly. The larger the stone the better, so enlist help to install them. In a pond lined with a flexible liner, use spare offcuts of liner or underlay where they will be mortared into position. After installation, regularly remove any algae from stepping surfaces with a wire brush.

WOOD

Wood can be used to make stepping stones provided every precaution is taken to remove slime and algae from the surface. Where it is possible to stabilize large log rounds in boggy soil, staple chicken wire over them to reduce the risk of slipping. Sections of round hardwood trunks, embedded in concrete to prevent them floating and to give extra stability, look effective and will last several years in shallow woodland ponds.

lighting

A pond is ideally suited to lighting, bringing a new dimension to its appeal as the source of light changes from natural to artificial. Since electricity may well have been connected to an existing pump, the main expense of lighting will have already been incurred. But as with all electrical appliances in or near water, any lights should be fitted by a qualified electrician using only approved fittings and connections. If the pool is viewed from a living-room window, make provision for an internal control for the lighting so that you enjoy the short-lived drama of hoar frosts, mists and even gentle rain during the winter months without having to venture outside.

Use subdued downlighters to illuminate the pool surround if the garden is used for barbecues and parties. If the lighting is extended well beyond the pool, the silhouettes and shadows are sharpened and made three-dimensional when the lights are positioned in both the foreground and background.

Rather than simply replacing daylight, artificial lighting creates the opportunity of projecting light from a different direction and angle, for instance from a spotlight hidden in a low position, to highlight a specific feature. Restraint and subtle positioning are the watchwords in successful lighting. Avoid using too many different colours and types of light. As technology advances, new techniques such as fibre optics and laser lighting will be used more frequently, but the

tips

- *Lenses are supplied in a variety of colours, but white is the most popular and more effective than a colour mix. Amber lenses are good for making the spouts of cobblestone fountains look like fires, but may bleach the colours of leaves.*
- *Try to subdue the external casings of light covers so that during daylight the stainless steel or white casings are hidden. Black is the ideal colour for light casings in the garden.*
- *Masks or gargoyles of wall fountains are more effective when lit by a narrow beam that highlights the mask but not the surrounds. Where the water from the mask falls into a reservoir pool, a submerged high-intensity, narrow beam pointing upwards into the spout makes it appear as if the light is travelling along the narrow thread of water.*

simple pleasure of seeing a small fountain spout highlighted from a favourite window still remains excellent value in lighting.

UNDERLIGHTING AND SPOTLIGHTING

Experiment with out-of-pool underlighting by using a powerful torch or a spotlight on a long lead. Choose your favourite viewing point, usually from armchair height, and pick out the best features or plants to underlight around the pool. Some plants that can appear rather dull in daylight take on a new lease of life with this treatment, and architectural plants such as gunneras, rheums and many waterside trees look stunning. Since there is so much more darkness in the winter, avoid highlighting plants that offer interest only in summer.

CREATING REFLECTIONS

Reflection has always been considered one of water's greatest attractions, and lighting allows the drama of plants, statues and urns to be reflected with enhanced effect in the inky black water. Tall or weeping specimens on the far side of a pool will create a mirror image provided the water is kept dark. Even ripples on the water surface create an interesting reflection. Again, use underlighting to highlight the object so that the light source is not reflected in the water. The water surface should be kept free of any direct light from above, and only use spotlights if they are submerged.

UNDERWATER LIGHTING

Like submerged swimming pool lights, underwater lighting adds a sense of mystery to deeper water. Illuminating a fountain or the turbulence under a waterfall so that the movement dominates the otherwise dark surroundings are the two most exciting options. It is less easy to experiment with underwater lighting before buying the actual lights, since insulation and safety precautions prevent any temporary experimentation. Safety rules now dictate that these lights are low voltage for domestic use. However, the increased light intensity of small halogen bulbs as compared to traditional tungsten bulbs does much to compensate for the lack of voltage.

Opposite: A coloured lens in the uplighter hidden in the cobbles gives this spouting wall mask added drama.
Above: Use narrow-beamed lenses and high-intensity submerged lights to highlight ornaments without bathing the surrounding wall in a bland light.

5 WILDLIFE

Any pond is a magnet to wildlife, but the more informal the pond, the more wildlife it will attract. By taking some simple measures in its design and planting, any decorative informal pool can become a wildlife haven. The main problem with formal ponds is the gap between the water surface and the surrounding paving, which amphibians find difficult to negotiate, especially when the paving has a slight overhang. However, formal ponds if properly planned and equipped offer the delights of ornamental fish-keeping, adding a whole new dimension of interest to water gardening.

creating a wildlife pool

An informal pool is similar to a wildlife pool, but to attract the maximum variety of wildlife, a number of requirements need to be incorporated into the siting and design of the pond.

SITING

The pool should be sited near existing vegetation to encourage shy creatures to visit, preferably in view of a window in your home. Make space allowance for a separate boggy area next to the pool, which will provide an intermediate habitat for worms and other invertebrates that would not survive if submerged. This enables a food supply to be at hand for frogs and toads.

DESIGN

The shape of the pool should look as natural as possible, avoiding fussy or contrived curves and long straight lines. It must be a minimum of 46cm (18in) deep, preferably 60cm (2ft) at its deepest point, with a variety of depths culminating in an area of the margin that is made into a shallow beach (see pages 58–9).

The bottom should be covered with a 10–15cm (4–6in) layer of soil to help overwinter many of the invertebrates in the mud, and there should also be an adequate selection of submerged water weeds for egg-laying and protection of the developing adult stages. Stones or rocks should be provided in the margins for hibernating amphibians to burrow under.

PLANTING

There should be an area of thickly planted vegetation around part of the margin. The planting should contain a large percentage of native species whose flowers act as a food supply for insects. Within the marginal planting there should be ample emergent leaves, such as iris leaves, which allow the larvae of dragonflies and other winged insects to climb out of the water and change to adults on the aerial parts of the leaves.

Left: Adult frogs can live in harmony with fish, but larger fish will soon devour tadpoles. To avoid this, you could make a shallow pool alongside the main pool as a useful nursery for frogspawn.

MANAGEMENT

Excessive hygiene should be avoided in and around the pool. For instance, dead leaves should be left on the plants throughout the winter (unless they fall into the water). This helps to provide protection and cover at a time of year when wildlife is at its most vulnerable.

INTRODUCING FISH

Highly decorative fish with bright markings, such as koi, not only look out of place in a wildlife pool but also have a marked, and probably detrimental, effect on the natural food chain, causing disturbance to the habitat at all depths of the pond. Koi are the main culprits in uprooting plants, and if a layer of mud has been deliberately introduced, this will constantly be disturbed and the water clouded.

The other major problem with large ornamental fish in the wildlife pool is their appetite for all sorts of submerged life, including tadpoles. When the fish are small (less than 8cm (3in) in length), their mouths are not large enough to swallow larger food and their impact on the food chain is limited. It is far better to introduce fish indigenous to the region, which will live in greater harmony with many of the other pool occupants.

Below: A mixed fish community is ornamental and useful in keeping down aquatic pests. Problems can occur as fish like koi grow to their maximum size and upset the balance in the pool.

fish-keeping

For many water gardeners, fish may be the main reason for having a water feature, and its design will have been influenced by this priority. For the majority of water gardeners, fish are the bonus, adding a new source of interest and providing movement and colour. Fish are also valuable in helping to control a number of pests. Orfe, for instance, devour any midge or mosquito larvae that alights on the pool surface. Ponds are a great attraction for these insects and fish will reward their keep by preventing an invasion of midges through nearby windows of the house. Fish also eat the larvae of pests that attack plants, such as the china mark moth and the water lily beetle.

STOCKING GUIDELINES

There are limitations on the number of fish that can be introduced in a pond. Fish release toxic ammonia as part of their body waste through the gill membranes, which could reach poisonous levels if too many fish are kept in a small pool. In a healthy well-balanced pool with ample plants, there are nitrifying bacteria present that feed on the harmful ammonia, converting it into beneficial nitrates which in turn are used by the submerged oxygenating plants. Pools that have a pumped filtration system can sustain a much larger stocking rate.

The recommended maximum stocking rate relates to the clear water surface of a still pond, that is excluding the area taken up by marginal plants around the edges. If the pool has been established for some time with mature plants and water has not been recently introduced, 5cm (2in) of body length of fish for every 0.09sq m (1sq ft) should be the top limit. Where a pond has been recently constructed and a minimum of 4–6 weeks has elapsed since the water and plants were introduced, stock with half of this density.

TESTING THE WATER

Problems for freshly introduced fish can occur in a new concrete pond if it has not been sealed or the water has been changed twice to remove the excess lime from the cement, which makes it highly alkaline. Fish are reasonably tolerant creatures and manage to adjust to most changes in their environment, but a rapid and severe change in the water chemistry is asking for trouble. There are simple kits available to test the water, which should be slightly acid or neutral.

Left: Large fish should be transported singly in shallow water inside oxygen-filled bags supported in a box. Two larger fish might damage each other in transit .

Fresh tap water is likely to be alkaline and after several weeks in the pool becomes more acid, particularly following heavy rain which is more acid than tap water.

BUYING FISH

It is better to buy fish that are 8–13cm (3–5in) long. If larger than this, they are not only more expensive but they find it that much harder to acclimatize to their new surroundings. Fish are particularly vulnerable just after introduction to the pond, and this is the time where they can easily damage themselves and risk fungal infection. Very small fish may not have the stamina to survive in a new environment, particularly in the autumn when they need build up body weight for the winter. A healthier balance of submerged pond life is achieved with several smaller fish than fewer larger ones, whose larger mouths are able to devour other important members of the food chain.

The ideal time to buy fish is late spring or summer when the water temperature is no lower than 10°C (50°F). Select fish with bright eyes and no evidence of damaged scales or white spots of fungal infection. The large fin on the back of the fish known as the dorsal fin should be erect and the fish should not be sluggish.

TRANSPORTING FISH

When the fish are to travel long distances or are likely to be in the transporting bag for some hours, take a suitable box in case the supplier gives you the fish in transparent bags. Place the bag in the box and close down the lid so that the fish travel in the dark. Good suppliers will have oxygen cylinders to inflate the bags with neat oxygen before sealing.

On reaching the pond, float the bag on the pond surface for about an hour to adjust the water temperature inside the bag to the surrounding pond temperature. After an hour the bag is opened and carefully unrolled to allow the pond water to spill into the bag. As the bag fills, the fish are allowed to swim free into the pond.

tip

- *When buying fish, look for more subdued coloration, which would suggest that they have been in quarantine for some time and are therefore less prone to disease on arrival in the pond. Highly coloured fish are generally indicative of freshly imported, more vulnerable stock.*

fish care

Above: Although they do not have the same variation in colours as koi carp, goldfish still remain the most reliable and adaptable fish for a coldwater pond.

Ensuring that the fish remain healthy requires a basic understanding of their needs, particularly since several species have long life spans. Koi, for instance, can live for 70 years. Space and depth requirements vary from species to species, with koi being the greatest demanders of room, requiring ideally a minimum volume of 10,000 litres (2200 gallons) and a minimum surface area of 20sq m (215sq ft). Since many ponds are considerably smaller than this, large fish such as koi should have supplementary filtration. Goldfish are more adaptable to small ponds but may not grow to their full size.

Apart from space to grow, another vital requirement for pond fish is adequate oxygen. This is absorbed from the atmosphere through the water surface and from the leaves of submerged oxygenating plants during daylight hours. Cold water is more able to absorb atmospheric oxygen than warm water, and in warm weather it is helpful to supplement oxygen levels by disturbing the water surface or introducing moving water through a fountain or waterfall. Since the oxygenating plants absorb oxygen themselves during the hours of darkness, there can be a crisis in midsummer in the early morning before dawn when oxygen levels

are really low. Dangerously low oxygen levels can also peak in summer after periods of prolonged sunshine, which causes a rapid growth of floating filamentous algae that in turn blocks out light necessary for the submerged oxygenating plants in the deeper water. Species of fish vary in their levels of tolerance of low oxygen.

Very cold weather may also present problems to fish. While most cold-water fish are able to survive winter in a torpid state on the pool bottom, shallow or raised pools that are exposed to prolonged frost will result in serious weakening of fish through absorption of toxic gases such as methane. This gas is formed through the decomposition of organic matter normally released into the atmosphere from the pool's surface, but in severe weather it is locked under the ice.

Rapid water-temperature fluctuations cause severe shock and topping up small pools in hot weather with large quantities of cold water from the mains supply can reduce the water temperature by several degrees. It only takes a drop of 4°C (7°F) within one hour during hot weather to kill spawn or fish fry.

tips

- *In a pool where there is adequate depth and surface area for both bottom and surface dwellers, it is a good idea to mix several species. Fish preferring the more oxygenated surface such as orfe will keep down midge larvae, while the bottom dwellers act as good scavengers.*
- *Quick-frozen 'live' food is available in small packs from specialist suppliers, which offers a welcome supplement to the everyday diet.*

FEEDING

This practice will have the most bearing on the ultimate health of ornamental fish. The larger the pool, containing ample nooks and crannies to harbour other submerged life, the less dependent the fish will be on supplementary feeding. Fish need a protein-rich diet and prefer to eat the eggs and larvae of other pond creatures. In a small pool, this source will soon be exhausted, particularly if there are high stocking levels. Maintaining an adequate supply of natural foods relies not only on the size of the pool but having ample planting and varied surfaces on which algae and crustaceans can thrive.

Without a varied natural diet, supplementary protein can be supplied in various forms such as freeze-dried, pelleted, flaked, floating and frozen. The main danger lies in overfeeding with one type of food, resulting in a build-up of uneaten and decomposing waste which encourages the spread of fungi. Floating food has great merit in allowing the food to be seen if uneaten. The guideline to the amount required is whether there is any left after 10–15 minutes of sprinkling on the surface. If possible, feed twice a day at the same time and in the same place. This will encourage the fish to become tame and even take food out of your hand. Live food provides a boost to the diet and fish suppliers sometimes have stocks of minute water life such as water fleas and cyclops.

Autumn and spring are the most important times to feed in order to build up reserves or replace body weight lost in the winter. Feeding should stop as soon as the temperature drops to around 10°C (50°F). Gradually reduce the food in colder weather by feeding with wheatgerm pellets rather than high-protein food. There are often mild spells in temperate winters when the fish may become active and appear hungry. Resist the temptation to feed in these circumstances since fish stomachs fail to digest food in low temperatures and it remains undigested in the gut. If you cannot resist the temptation to feed, only use wheatgerm food.

BREEDING

Healthy mature fish should have no difficulty breeding in a well-balanced pond providing they have reached adequate size, which is around 13cm (5in) for goldfish and 25–30cm (10–12in) for koi and orfe. To prevent cannibalism, there must a sufficient amount of oxygenating plants with fine foliage in which the spawn is lodged and the fish fry hatch and hide.

Opposite: In summer water lily leaves protect fish from herons and provide shade. Cover the roots with large cobbles to stop larger fish disturbing them. *Above*: The water lilies' combination of foliage shape and flower colour is perfectly complemented by the movement of fish in this dark water.

CHOOSING FISH

In the same way that the choice of plants relates to the type and design of pond, so certain fish species are more appropriate for particular types of pool. Wildlife pools, as we have seen, are better stocked with native fish, while the design of an ornamental pool will influence the colours and size of its inhabitants. The pool's depth also has a strong bearing on choice. A shallow pool would be totally unsuited to koi, which require deep water of 1m (3ft) or more, but such depth would be unnecessary for fish such as orfe which prefer shallower water. Climate is another influencing factor – even relatively tough breeds such as goldfish have variants that do not thrive in low water temperatures.

The fish featured here represent the more common, tried-and-trusted species that should present few problems for the water gardener. For the more adventurous fish-keeper, the fancy forms of goldfish, such as lionheads, orandas and celestials, are a sound choice. With the increased interest in wildlife pools, there are also many more species of native fish available. These are not easily seen in a pool because of their habits and colourings, especially in one that is heavily planted. The more exotic fish, while being easier to see, are more vulnerable to predators and appropriate measures should be considered to deter them, particularly cats and herons.

Above left The colours and markings of koi carp are undeniably attractive, especially when they dart around at feeding time.
Above: Orfe enjoy the highly oxygenated water at the base of a waterfall. Elegant fish seen from above, they look good in shoals.

FANTAIL GOLDFISH

The common goldfish, which live for about ten years on average, are usually orange-red in colour and feed at all levels. They grow to around 15cm (6in) and are extremely tolerant of a wide range of temperatures. The goldfish or shubunkin (calico) fantails make an attractive addition to the fish selection with their unusual long, double caudal fin and a more rotund body than the common goldfish or shubunkin. The goldfish fantails are referred to as red or red and white fantails, whereas the shubunkin fantails are referred to as calico fantails. Their shape and fin slow them down in the water, and although this makes them easier to watch, they are extremely vulnerable to predators. Goldfish fantails can turn almost white when mature, particularly when fed on a high-protein diet. They seldom exceed 20cm (8in) and are less hardy than common goldfish. In autumn, when the temperature drops below 15°C (59°F), they are best moved out of the pool and into an unheated aquarium until the spring.

KOI

Koi are all the same species and are named according to their colouring, which originates from black, white, red and yellow colour cells. Since there may be several colours in one fish, some of their names can build up to three or four words. They are not suitable for small ponds because they can grow up to 60cm (24in) in length. Due to their voracious appetites and the fact that they often eat plant material, the large specimens need an efficient filtration system to deal with their waste. They feed at all levels and prefer temperatures of 4–20°C (39–68°F).

GOLDEN ORFE

A European native, this decorative fish has an orange back covered with scales of a golden sheen. It is a very active fish and provides a good contrast with slower-moving fish such as goldfish and koi. It spends much of its time just under the water surface looking for insects. It is a naturally shoaling fish, so it is a good idea to have a minimum of four and allow them adequate room to move – at least 4.6sq m (50sq ft). They prefer well-oxygenated water and would be one of the first to suffer during hot nights in summer. They also dislike long periods of cold weather and may suffer if the water temperature persists at 1–2°C (34–36°F). For this reason, it is advisable not to disturb the warmer water that accumulates at the pool bottom during long cold spells. They do not disturb the bottom like goldfish and koi, and feed on zooplankton, aquatic insects, fish fry, tadpoles, algae and some water weeds. They can grow to 30–50cm (12–20in) with a life span of 15–20 years, reaching breeding age at about three years. Young orfe can be distinguished from goldfish by their short dorsal fin.

Right: As well as stealing fish, herons can accidentally pierce flexible liners with their dagger-like beaks.

fish pests and diseases

FISH PESTS

Heron

By far the most serious problem for the ornamental fish-keeper is the loss of fish to the heron. Herons are long-lived birds which obtain their food by wading in wetlands and visiting garden ponds. They can become a common sight by the side of water, where they stand on one leg with eyes appearing half closed and head hunched into the tops of their wings. Without warning, they pounce onto an unsuspecting victim in the shallow water with their dagger-like beaks. Once they identify a rich food source of ornamental pond fish, they persist in hunting them relentlessly and with great guile. With many of the common remedies, including decoys, proving ineffective, control is not easy. There is a proprietary heron guard that relies on the heron pushing an inconspicuous cord and releasing tension on a sprung device, setting off a small cap. If this device or cords around the pond fail, you may have to resort to laying netting (plastic mesh) over the surface of the pond.

Water boatman

This is one of the most common species of water bug, which can detect the slightest movement on the water surface. Consequently, very few small creatures escape its attack. Despite its relatively small size of 1.7cm (⅝in), it is capable of killing a small fish by injecting poison through its piercing mouthparts. It resembles a

Left: Although adult dragonflies are beautiful the nymphs are a nuisance as they eat small fish.
Right: Feeding time provides a good opportunity for checking that fish are healthy.

miniature boat, with its wings making the hull and the oars formed by a third pair of legs. It is sometimes called the 'backswimmer' because of its habit of swimming upside down. Control is not easy since the adults can fly from pond to pond on summer evenings. If young fish are being killed in numbers, then netting the pool is the only method of control.

Water scorpion

These creatures resemble scorpions, although they are unrelated to them, and are about 2.5cm (1in) in length. They live in shallow water and move very little. Being poor swimmers, they prefer to sit and wait for their prey to come to them. When a small fish passes by, the water scorpion grabs the unfortunate victim using its front pincer-like legs, which hold on while the sharp mouthparts pierce the fish and kill it. It is able to breathe by its sting-like spine at the tip of its flat oval body, which acts like a snorkel. The only control possible is by rigid hygiene within the pool and hand-plucking when the small oval tubes are seen on the water surface.

Great diving beetle

Spending most of its three-year life span in water, this common large beetle is distinguished from other insects by its hardened wing cases, which act as protection for its delicate hind wings. It inflicts a ferocious bite and is capable of killing small fish, in addition to newts and tadpoles. It reaches 5cm (2in) long and the hard body is dark brown with a distinctive yellow or golden edge. Its larvae resemble dragonfly larvae but are smaller, reaching only 4cm (1½in) long. They fly at night, enabling them to spread to other ponds with ease, landing on new ponds and swimming on the surface with their flattened and elongated hind limbs. In order to breathe, the larvae float to the surface tail first and stick their tails out of the water to draw air into their breathing tubes. The adults store air in their wing cases. Control is difficult, limited to catching the adults in a net when seen.

Dragonfly larvae

These ugly creatures can spend up to five years as nymphs underwater, turning into more attractive adults for a life span of 1–2 months. It is the scorpion-like nymphs that predate on small fish, spending most of their time waiting on the muddy pool bottom. They catch their prey by shooting out their 'mask', which is an extension of the lower jaw and resembles lobster claws. They have no pupal stage and when ready to emerge as adults they climb up plant stems, shed their skins, dry out their wings and fly away as adults. Control is difficult, requiring hand removal of the nymphs when spotted.

Fish parasites

Several minute parasites attack fish causing symptoms of lethargy, loss of appetite, loss of weight and rubbing against the pond bottom. Fin damage may occur and infected fish secrete mucus in such quantities that their respiratory functions are reduced. Most parasites are invisible to the naked eye but the largest are anchor worms, fish leech and fish louse. Control is through the use of proprietary products and antiseptics.

FISH DISEASES

Fungus or 'cotton wool disease'

This is one of the most common diseases attacking all types of fish. Fish are more susceptible when they have been weakened by stress, injury or infection. Cotton wool-like growth appears on nearly all parts, but in particular where damage may have already been caused. Several remedies are available from the fish supplier.

White spot

This is another fairly common disease, particularly in newly purchased fish. As the name suggests, several white spots appear all over the body resembling grains of salt. The fish rub themselves on the pond bottom to relieve the itchiness caused by the spots. Proprietary remedies are available for treating the whole pond.

Fin or body rot

This mainly affects fish with large ornamental fins such as goldfish and shubunkins. The diseased area becomes ragged and bloodshot, and is frequently affected by a secondary infection of fungal disease. The fish needs to be caught in a soft net and the affected parts of the fin removed with sharp scissors. Treat the cut parts afterwards with a fungicide such as methylene blue.

6 PLANTING

Aquatic plants have two roles in the pond – they keep the water clear and enhance the pool through the beauty and form of leaves and flowers. Designing a planting scheme that serves both these functions involves a careful balancing of various types of aquatic plant, from small and inconspicuous submerged plants to the more showy larger plants around the margins. Pond plants never suffer from a lack of water and as a result grow very quickly. Therefore, avoid the temptation to use too many plants and take account of their ultimate size in relation to the size of the pond.

aquatic plant sources

Since the majority of aquatic plants available are native plants, it could be tempting to scavenge local natural ponds for plants to save buying them. There are very good reasons for not doing this, the most important being that it is illegal to dig up certain plants. Natural wetlands, which are becoming increasingly scarce, provide homes for a diverse variety of wildlife, and the removal of plants puts the survival of more and more species in jeopardy.

Wild aquatic plants are invariably covered with snail eggs, algae and often the tiny leaves of duckweed, all of which spread with alarming speed when introduced into a new pond where there is little competition. It is also likely that the plants removed will be totally unsuitable for a small pond and may even damage a flexible liner. A typical example of this unsuitability is the reedmace (*Typha*), sometimes referred to as bulrush. This fast-spreading plant has dagger-sharp tips to its roots and these can pierce liners very easily.

Plants grown in a nursery will have been raised from seed or through techniques that pose no threat to endangered plants and are much less likely to introduce harmful organisms to the new pond. The labels on the plants will often give details of the flowering time, the optimum depth of water for planting and the ultimate spread and height of the plant.

planting schemes

Priority in any planting must be given to the functional plants that help to keep the water healthy and clear. A pool without plants will quickly go green in sunlight through the growth of minute, single-celled green algae. These algae feed on the mineral salts in the water, and a new pool that has been recently filled with tap water is highly charged with these salts. The remedy is to use two types of plant – the oxygenators, which feed on the mineral salts and starve out the offending organisms, and the surface-leaved plants, which shade out the light and further weaken the algae.

Once the functional plants have been chosen, with many aquatic plants offering dramatic shapes and lush growth, there is no shortage of opportunity to be creative in planting, both in the selection of single architectural specimens and the grouping of informal combinations. Water gardening also offers an additional and unique design dimension – reflection.

Many of the rampant aquatics can still be grown in small and medium-sized pools by planting in aquatic planting baskets and frequently dividing them, allowing a greater variety of plants to be used in a confined space. Making full use of leaf shape and colour is important, since the size and colour of flowers are not significant features in most temperate aquatic plants. Visits to nurseries, flower shows and gardens are invaluable in compiling your own list of favourites, noting how they look throughout the seasons.

Keeping interest in the plants throughout the year is not easy in a small pond because most of the aquatics die down, leaving straw-coloured, withered leaves for the winter. Plant interest in the winter can be enhanced if the surrounding area is included in a planting plan. Winter is one of the best times to enjoy reflections on the water after the water lily leaves have died down, and coloured stems and the outlines of adjacent plants become highlighted in the water.

Formal and informal ponds will require different styles of planting, both in the deep water and in the margins. Water lilies are common to both styles, but colour is given greater emphasis in formal schemes.

PLANTING FORMAL PONDS

The simplicity and symmetry associated with formal design should be reflected in the planting, using bolder groups of fewer species rather than too much variety. Formal schemes make no attempt to be 'natural', so are best suited to the use of bright colours. Formal planting can also make much more use of single specimens, particularly in the deeper water where the plant's impact will be greater surrounded by clear water.

PLANTING INFORMAL PONDS

There is more freedom in the choice of plants for informal ponds, but this can result in a tangled mix of uncoordinated and overgrown plants. Shape and more subdued colours provide summer splendour rather than strong colours. Shapes should not compete and adjacent plants

tips

- *To make the maximum use of the pond's reflective surface, plant a striking plant on the far side of the main viewing area, limiting the nearside edge to minimal or low planting.*
- *To create the illusion of a larger pond from one viewing position, plant a large architectural plant on the near side, then a smaller one with a similar shape on the far side. This increases the perspective and gives a feeling of greater size.*

Left: This formal canal leading to a circular pool has been converted into a moist bed for globeflower (*Trollius*) and iris.
Right: A mixture of heights, textures and colours crowd round this pebble-edged pool, creating a lively atmosphere.

are better if they are contrasting in nature. For example, a creeping plant makes a better partner for a tall plant. Vary the planting density around informal pond edges. A pool with an even ribbon of plants is less effective than one with mixed group sizes, showing heavy planting in one area and little in another. This gives easy access to the pool in some parts and almost impenetrable areas in others.

planting guidelines

Prepare a planting plan before buying any plants. Draw the area on graph paper to make planning to scale easier, and mark the area that each plant will occupy with a circle. Start with the deeper water where the first choice is either a mixed colour range or a single colour theme of water lilies. It is tempting to fill the pond with as many water lilies as possible, but the effect is much better if the leaves only cover between a third and half of the pond's surface.

Before further selection, decide if there is one time in particular when you want any flower colour to peak or whether you want to spread the flowering season out. For the rest of the plan, choose and plot in the tallest species first, then infill with the creeping plants. Bear in mind that some marginal plants can reach nearly 2m (6ft) and would be out of proportion in a small pool. As a rough guide, limit the height of the taller plants to no more than two-thirds of the length of the pool. The taller plants are better sited at the appropriate end of the pond to prevent their shade falling on the water. Generally avoid grouping plants with a similar shape and height next to each other.

Where pond sizes are larger than 5.5sq m (60sq ft), it is advisable to plant the marginals or moisture-lovers in groups of three or five rather than singly. When planting favourite groups or specimens, insert a tall cane in the proposed area and view the site from your favourite windows, particularly in the kitchen or living room. Keep moving the cane until you are completely satisfied that the proposed planting can easily be seen.

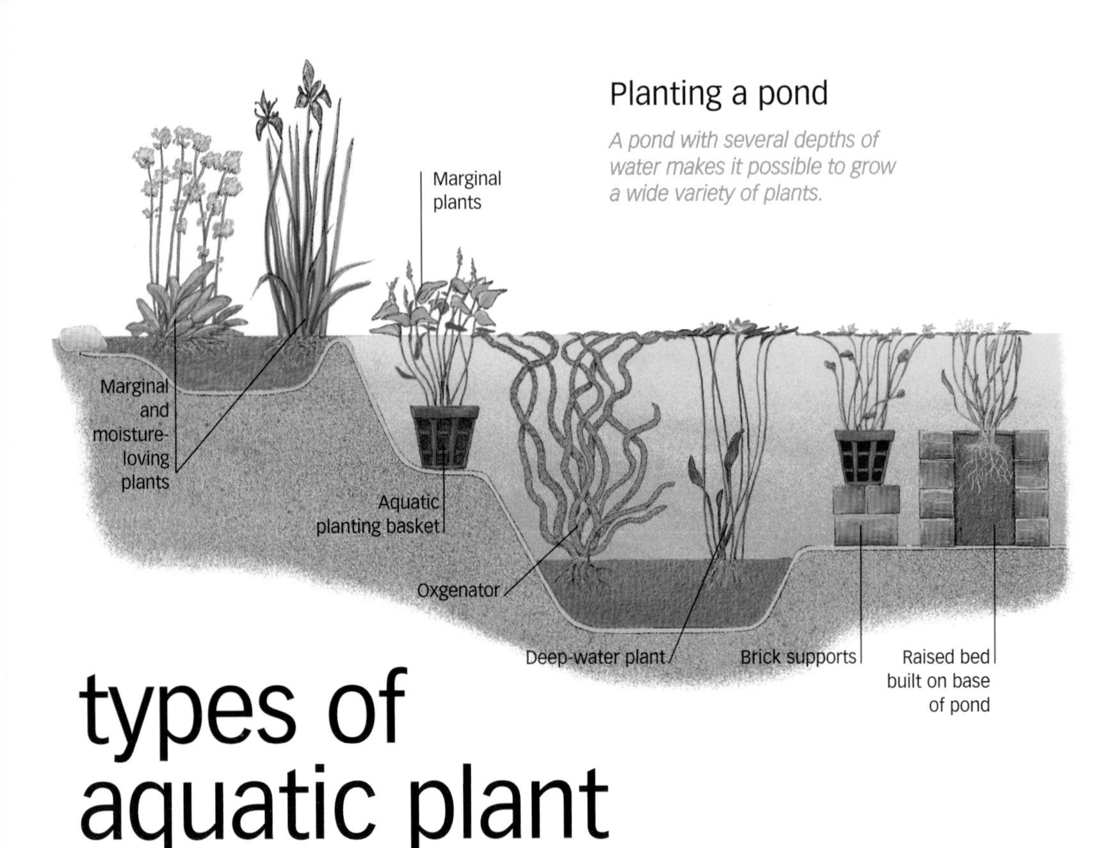

Planting a pond

A pond with several depths of water makes it possible to grow a wide variety of plants.

types of aquatic plant

Water plants are categorized according to their planting environments, as follows: submerged, generally known as oxygenators, floating, deep-water and marginal plants.

OXYGENATORS

These submerged plants are the workhorses of the pond, generating oxygen in the daytime from their leaves and feeding on dissolved mineral salts which helps to keep the water clear. Most oxygenators provide shelter and food for the minute aquatic animals including fish fry.

Many oxygenators have exquisite and delicate foliage because it is not dependent on fibrous tissue to maintain rigidity. The leaves, being suspended in the water, are often thin and translucent. Certain species develop attractive whorled arrangements of leaves, such as the milfoils or parrot feather (*Myriophyllum*). Being dependent on water chemistry through their extensive thin leaf surfaces, it is advisable to include several different oxygenators in any pond planting to ensure that at least some of them survive if extremes of light, temperature, acidity or alkalinity are experienced at any time.

FLOATING PLANTS

One of the most common and typical of the floating plants, which spreads across both temperate and tropical locations, is duckweed (*Lemna*). Although regarded as something of a pest in the ornamental pond, it exhibits an extraordinary ability for rapid reproduction, generally by offsets, which is the typical method for this group of plants.

Floaters are extremely useful in the establishment of a new pond as they shade the surface until the larger leaves of the water lilies and other species begin to cover the surface. This helps to prevent the water greening in the new pond by reducing light and mineral salts in the freshly installed water. Often a covering of duckweed hides crystal-clear water underneath. On a small pond, the floating plants can be netted off at a later stage when they completely cover the surface or when there are enough other leaves on the surface. On larger ponds, however, they must be used with great caution since later control could be extremely difficult.

Newly introduced fish also appreciate this protective surface cover, particularly in a new pond where the oxygenators have not

Right: Despite the value of duckweed in shading a newly established pond, it becomes difficult to control once it spreads between marginals and must be thinned regularly.

yet fully developed. To compensate for their lack of anchoring, many floaters have extensive and fine root systems with exposed root hairs. These offer an additional bonus to the shade their leaves provide in creating a refuge for the fish fry. When viewed in an aquarium, the roots of floating plants such as water lettuce (*Pistia stratiotes*) and water hyacinth (*Eichhornia crassipes*) make an attractive feature. All floating plants prefer nutrient-rich ponds.

DEEP-WATER PLANTS

These plants grow in the deeper water, usually between 30cm (12in) and 60cm (24in), and are best illustrated by the water lily. Firmly anchored to the bottom, deep-water plants fulfil both aesthetic and functional roles in providing flowers and shade. They offer resting platforms for a variety of creatures and a depository for eggs on the undersides. Their leaves prefer still water and unlike totally submerged plants are less tolerant of currents. Most survive temperate winters in the form of thick rhizomes under the ice.

MARGINAL PLANTS

As their name implies, these plants prefer the shallower conditions of the pool margins and generally have a tolerance of water up to 15cm (6in) above their crowns. Their leaves tend to protrude from the water rather than float on the surface. Many of them will grow without water over their crowns and tolerate a degree of drying-out in the summer, but would not permanently survive in these condition. The reeds and rushes best illustrate this group, which remorselessly invade natural ponds from the margins.

Marginals give the height to a pond planting scheme typically with sword-like leaves being dominant. Some marginals have curious spathe-like flowers, which in the case of the skunk cabbages (*Lysichiton*) are an unforgettable sight when planted in groups alongside a stream, the flowers being followed by large paddle-like leaves.

Below: White skunk cabbage, *Lysichiton camtschatcensis* likes humus-rich damp soil.

Right: To keep plants, such as this *Elodea crispa*, in submerged baskets, add a layer of pea gravel as a top dressing.

planting techniques

Except for small wildlife ponds where the plants are planted directly into a layer of mud on the pond bottom, most decorative ponds are best planted with containers. These allow the plants to be moved easily in the pond, and when the plants need attention, it is a relatively simple job to remove the container from the pond rather than dig the plant out of a permanent soil bed under the water. Plants in containers are much less likely to become overgrown and swamp neighbouring plants, as can happen in ponds with a soil base.

A good compromise between container planting and planting into the soil base of a wildlife pool is to build permanent raised beds on the base of the pool and around the edges during construction. These are filled and planted before the pond is filled with water, making it possible to achieve a natural appearance while at the same time controlling the plants' spread.

tip

- *When submerging the containers in deep water, thread string through the sides to form handles. This makes it much easier to position the crate as it is lowered into the mud.*

CONTAINER PLANTING

Aquatic containers come in all shapes and sizes, including curved containers to fit the ledges of informal ponds. All have a wide base to give rigidity – an important requirement for tall marginal plants that are easily blown over in high winds.

The containers have mesh sides to allow gases and solutions to move easily into the pond water and prevent the soil from stagnating inside. In order to prevent the soil leaching out, the containers are lined with hessian squares or squares of finely woven polypropylene sheeting, both of which are permeable enough to allow the easy interchange of gases and water. Containers with very fine louvred mesh sides eliminate the need for lining.

Aquatic containers range from 40cm (16in) in diameter with a 36 litre (8 gallon) capacity

Right: *Orontium aquaticum* (Golden club) prefers deep mud in water no more than 45cm (18 in) deep.

suitable for medium to large water lilies, down to 4cm (1½in) in diameter with a 50ml (2fl oz) capacity for aquarium planting.

PLANTING MEDIUM

Aquatics will grow quite happily in a garden soil that is preferably on the heavy side and has not been recently manured or fertilized. Loose organic matter should be sieved out since this will float, and avoid very sandy soil that contains little nutrient.

Proprietary composts used for terrestrial plants should not be used for aquatics because they contain too much peat, which will eventually float. They also contain too much fertilizer which will dissolve in the water and cause it to go green. An ideal source of suitable material is an old turf heap in which turves have been allowed to rot for a few months. Where there is no suitable garden soil available, proprietary aquatic composts are available from specialist suppliers.

PLANTING TIME

Unlike the majority of terrestrial plants, aquatics should be planted while in active growth, preferably between spring and late summer when water temperatures are warm and there is plenty sunshine to stimulate growth. If planted too late in the year, the plants have insufficient time to become established in the new soil before they die down for the winter. This is particularly important for water lilies, which must build up food reserves in their roots before winter if they are to survive and flourish the following season.

PLANTING DEPTHS

The ideal planting depth in the water varies from plant to plant and there are guidelines given on the labels of plants from most good suppliers. Where this depth is given, it is the measurement from the top of the soil in the aquatic container to the water surface. Never plant too deeply; without adequate sunshine the plants will become starved and die. Initial planting may require standing the container on bricks or blocks so that the plants are not too deep when they are young. As the plants grow they can be gradually lowered to their optimum depth by taking away the blocks underneath.

PLANTING FLOATING PLANTS

Since these plants have no anchorage, they are simply scattered on the surface of the

Left: The young leaves of water lilies, for example this *Nymphaea* 'Escarboucle', must be able to reach the surface as soon as they are planted. Achieve this either by raising the container on bricks and lowering it over a period of weeks, or by gradually raising the water level.

pond. The initial positioning is unimportant because the groups of plants are moved around on the surface by the wind and so will look natural.

PLANTING OXYGENATORS

There are normally about six species of oxygenator to choose from at good aquatic centres, so buy a good mixture. They are often sold in bunches of unrooted stems and a medium-sized planting crate will hold five bunches, one in the centre and one in each corner. Keep plants of the same species together in a container so that the vigorous ones do not overrun the more delicate ones. As a rough guideline, plant 5 bunches to every 1sq m (10sq ft) of surface area. Submerge the cuttings in the pond as soon as possible after planting to prevent them wilting and shrivelling.

If you have a mud-lined pond and you have bought the plants in bunches with a lead strip around the base of the stems, they can simply be thrown into the pond where they will sink and root in the mud.

Oxygenators are increasingly being sold in small fibre cubes that disintegrate with time. These can be thrown into the pond and will provide sufficient sustenance and anchorage to allow the oxygenator to become self-supporting in a few months.

PLANTING DEEP-WATER AND MARGINAL PLANTS IN CONTAINERS

Settle the plant into an appropriate compost in a container large enough to fit the plant when fully grown. Line large mesh containers with a square of hessian or polypropylene before filling up with compost.

After inserting the roots into the compost, firm the soil around the roots and leave a gap of 2.5cm (1in) between the top of the compost and the rim of the container. Top-dress the compost with washed pea gravel to just below the rim to avoid losing compost when immersed.

PLANTING WATER LILIES

The roots of water lilies fall into two different types, and this determines how they are planted. The rhizome-type root should be planted horizontally just under the surface of the compost with the crown just protruding from the surface. The thicker tuber-shaped roots, with their fringe of fibrous roots just under the crown, should be planted vertically in the compost, again with the top of the crown just protruding. Ensure that the compost is really well firmed before top-dressing with gravel.

Submerge only to a depth where the young leaves can reach the surface immediately. This is achieved by supporting the container with blocks so that there is no more than 15–20cm (6–8in) of water above the crate. After two or three weeks, growth should be strong enough to lower the container gradually, one brick's depth at a time, to the bottom of the pond. If vigorous water lilies are planted into permanent box-like raised beds on the pond bottom, the water level should be kept just above their growth and more water added in stages until the pond is filled.

Above: Iris laevigata flowers in late spring to early summer. It can be cut back in late summer.

pruning

The rapid growth produced by aquatic plants can cause ponds to become overgrown very quickly. Although the risk of swamping the pool with vegetation is lessened by restricting plants in aquatic containers, it is still necessary to cut them back regularly, particularly in mid- to late summer when the volume of growth is at its peak.

OXYGENATORS

Submerged plants are the greatest culprits for overgrowing ponds, since they can soon escape from their containers and grow successfully without any anchorage. Oxygenators quickly reach the water surface, and if all the summer's growth is allowed to die back throughout the winter months, masses of decomposing vegetation is produced which denies the water oxygen and produces harmful methane. Autumn is a good time to give the long submerged stems of oxygenators a hard prune, since the need for their services decreases in the winter with lower light and easier absorption of atmospheric oxygen through the cooler water surface.

WATER LILIES

Water lilies also need to be pruned back periodically. Both the flowers and the leaves are being constantly renewed, so they should be removed once they have faded or turned yellow. In large pools where the plants are hard to reach, a long-arm pruner is useful to save wading in the water.

MARGINALS

The vigorous species of marginals also benefit from a regular cutting back from mid- to late summer, especially those that have flowered earlier in the season, such as many of the rushes. They quickly regrow and the younger shoots are healthier and offer less wind resistance. A complete cut-back of the marginal foliage is more appropriate for the clean lines of a formal layout, whereas the dying leaves should be left around the wildlife pool until late spring to give protection and cover for small creatures. Once the young growth starts shooting in spring, this is a good time to give marginal plants a really thorough clean-up.

FLOATERS

The floating plants that may have been introduced in the early establishment of the pond have a habit of spreading rapidly and covering too much of the water surface. Keep an eye in particular on the spread of duckweed (*Lemna*), fairy moss (*Azolla filiculoides*) and water soldier (*Stratiotes aloides*), and remove with a net to keep in check. Some of the floaters are not frost hardy, such as water lettuce (*Pistia stratiotes*) and water hyacinth (*Eichhornia crassipes*), so these plants are unlikely to become a problem in temperate climates. They need to be removed to a frost-free environment in order to survive the winter.

Left: Water hyacinth (*Eichhornia crassipes*) can be divided at any time of year by taking offshoots. It is not frost-hardy, so it must be overwintered indoors.

division

Splitting and repotting of plants needs to be carried out regularly in the pond since their rapid growth soon exhausts the food supply and the plants become weak and straggly. Unlike terrestrial plants, aquatics cannot be fed by the liberal spreading of fertilizer around the pond. This will only result in rapid greening of the water.

MARGINALS

Plants in containers should be divided regularly – they become difficult to tease apart if they are left for too long and the roots grow through the mesh of the containers. Where this has happened, it is best to sacrifice the container by chopping it up with a sharp spade, thus allowing the new growth around the sides to be released. Discard the older shoots in the centre in favour of the young growth around the sides and repot the new shoots into new containers using fresh compost.

WATER LILIES

If water lily roots escape from the container, they can quickly spread across the pond, growing fine root hairs into the thin layer of mud on the bottom. A sign that they need dividing is a mass of leaves at the expense of flowers, growing upright above the water surface instead of sitting on the water. All of the old thick root should be discarded and the tips of the young growth then planted into fresh compost.

OXYGENATORS

Since oxygenators are not so dependent on an extensive root system as other aquatic plants, there is no need to divide them. Take cuttings of the young shoot tips instead. The cuttings need to be about 30cm (12in) long and bunched together before repotting or weighting down with a small piece of lead, enabling them to sink to the bottom. The old shoots and container should be removed from the bottom of the pool and discarded.

feeding

The only plants likely to require any form of supplementary feeding are the water lilies, particularly the smaller cultivars. To avoid the danger of excess fertilizer causing greening of the water, special slow-release tablets are available that are simply pushed into the compost of the water-lily container.

Right: Green leafhoppers (*Cicadella viridis*) attack the undersides of water lily leaves, if they do not sit flat on the water surface.

plant pests and diseases

Pests and diseases, although relatively few compared to those afflicting food crops and intensively grown ornamentals, can spoil the appearance of water plants, particularly in a small pond where they are seen at close quarters. Pests are more of a nuisance than diseases, mainly attacking water lilies. Good cultivation reduces the problems considerably, since weak plants in need of dividing are more prone to attack than young healthy ones.

Biological control as opposed to chemical control is vital in the water garden if fish are present, and even a pool without fish contains considerable life that would be killed by chemicals dissolved in the water. Fish are great allies in the war against pests. Where small insects are eating foliage beyond the reach of the fish, spray the leaves with a powerful water spray to dislodge them onto the water surface to be devoured. If an infestation or attack becomes so severe as to require chemical control measures, remove the plant from the pond and immerse in a separate container filled with the chemical for a few hours. Wash the plant thoroughly in clean water before returning to the pond.

PLANT PESTS

Aphids

The soft leaves of pond plants prove a great attraction to aphids, particularly the water lily aphid which attacks tissue above the water line. Fish will soon devour any they can reach, as do ladybird larvae, underlining the importance of using biological control in the water garden. In late summer the water lily aphids leave the pool to overwinter on the stems of nearby plums and cherries, and if these two fruits are sprayed with winter wash to kill the eggs, the life cycle is destroyed and the population of aphids reduced.

Leafhoppers

These are also attracted to the leaves of water lilies, particularly the crowded leaves of overgrown plants that tend to stick up in the air. Like aphids, they damage the plant by sucking the sap and in severe infestations cause the leaves to go brown. They are unlikely to be a problem if the leaves sit flat on the surface and the crowns are regularly divided.

Brown china mark moth

While the leaf-sucking insects such as aphids and leafhoppers reduce the vigour of plants, they are not nearly so obvious a pest as the moths whose larvae bite into leaves, causing unsightly holes and in severe cases skeletonizing leaves. This moth is one of the most common whose larvae eat large holes around the leaf margins.

Left: Caddis fly (*Halesus radiatus*). The larvae protect themselves underwater in a case of sand, pieces of shell, sticks and leaves while they chew foliage, such as water-lily leaves.

These small moths are about 2.5cm (1in) long, have irregular white patches on brownish-orange wings and are generally to be seen during the evening in the latter part of the summer. Their creamy-coloured caterpillars hatch from egg clusters just beneath the surface of the leaf, and after chewing up the foliage protect themselves before pupating by forming shelters out of cut pieces of foliage which they stick onto the leaf. The caterpillar is then able to crawl about and feed with only its head and forelegs protruding, giving it the common name of bagman caterpillar. Control is by removing affected leaves and netting any pieces of floating foliage acting as homes for the over-wintering pupae.

Caddis fly

Another insect that resorts to protection inside a case for part of its life cycle is the caddis fly. The adults resemble small dull-coloured moths with greyish or brown wings, seen struggling in flight in early evening when they lay nearly 100 eggs in one batch in or around water. The eggs are laid in long, jelly-like tubes sometimes seen dangling from leaves on or near the water surface. After emerging from the eggs in about 10 days, the larvae immediately start to spin protective cases made of small sticks and pieces of shell and sand, which are covered with fragments of leaves. They then swim or float around in their structures, feeding greedily on any vegetation present and often desiccating water lily leaves. They later pupate in leaves at the water's edge. Carp, golden orfe and goldfish are their natural enemies; chemical control is almost impossible.

Leaf-mining midge

This is one of the most disfiguring pests to attack water lilies. After hatching from eggs laid on leaf surfaces, tiny larvae eat serpentine channels through the surface of floating leaves. Eventually the damaged tissue dies away, resulting in large areas of skeletonized tissue. If control by removing affected leaves is not enough, remove the plant and immerse it for an hour or two in a bucket of dilute insecticide that is used for killing aphids. Carefully wash the plant before returning it to the pond.

Water lily beetle

This less common grub is one of the most destructive of water lily pests. A mere 0.6cm (¼in) long, this dark brown beetle lays its eggs in clusters on the surface of water lily leaves in early summer. The eggs hatch out in about seven days into shiny black larvae with distinctive yellow bellies. The larvae feed on most vegetation, stripping away the surface tissue and reducing the leaves to a skeletal framework that gradually rots away. The larvae pupate on any aerial foliage and the resulting beetles overwinter in the brown foliage at the water's edge.

Removing this dying foliage in the autumn is an important control measure. During the summer, early removal of infected foliage and spraying the leaves with a powerful water spray helps to check the spread. Fish devour all stages of the pest's life cycle, so great caution is advised on any control measure involving a pesticide. In addition to dislodging the larvae from the leaves by spraying, the lily pads can be temporarily submerged by raising the water level or placing wire mesh over them to submerge the pests for the fish to eat.

Snails

Snails should generally be regarded as a pest rather than an ally. The exception to this is the ramshorn snail, a good algae eater and easily identified by its rounded shell in the shape of a

ram's horn. The most common snail found in ponds is the great pond snail or freshwater whelk, which acts as an intermediate host for several fish parasites. It is almost impossible to prevent its introduction since the jelly-like eggs arrive unnoticed on plant material bought at aquatic centres. The snails are identified by their tall, pointed and spiralled shell about 2.5cm (1in) high. They like soft tissue and devour young water lily leaves with relish. They can be trapped by floating a lettuce leaf on the water surface for 24 hours and then scooping them off with a net.

The great pond snail (*Linnaea stagnalis*) is particularly fond of young water lily leaves, as well as eating the soft tissues of other plants.

PLANT DISEASES

Leaf spot

Water lilies are the most likely water garden plants to encounter any disease and in most cases this takes the form of a leaf spot. Leaf spot appears as either dark patches or as dry brown edges. The dark patches on the leaves eventually rot away, leaving an almost disintegrated leaf. If fish are present, the best remedy is to remove infected or dying leaves at the first sign of infection or yellowing. In the absence of fish, weak fungicidal sprays can be tried.

Water lily crown rot

This is a serious but less common disease that should not occur if the plants are bought from a reputable source. The leaves turn yellow, sometimes mottled, and the leaf stems become soft and blackened. The leaves finally break away from the crown, which is softened and gelatinous with a blackened internal tissue that smells foul. There is no control suitable for the garden pond other than quickly removing the affected plant, since the disease can pass to other water lilies through the water. If caught early enough, there is a chance that other water lilies in the pool will not have been affected, particularly if they are different varieties or species to the affected plant. If all seems clear after a few weeks, the removed plant can be replaced with a new one.

Iris leaf spot

Iris leaves suffer from a particularly disfiguring form of leaf spot. Brown oval spots appear randomly, later becoming elongated in the direction of the veins. The centre of the spot is often lighter and may show the spores of the fungus. In very wet conditions, the spots may coalesce before the leaf finally dies. A broad spectrum fungicide such as dithiocarbamate may be used provided there is no spray drift onto the water.

Mildew

The heavy canopy of surrounding lush foliage and the damp atmosphere at the water's edge make ideal conditions for mildew to flourish, but fortunately it is mainly restricted to marsh marigolds (*Caltha palustris*) in mid- to late summer. Mildew is unlikely to kill the plant, and unless the appearance of the grey coating of mildew offends, it can be left alone. If the mildew becomes too obtrusive, remove the diseased foliage to encourage the rapid growth of fresh young leaves.

7 PLANTS

A pond is as attractive to the keen plantsperson as it is to the fish-keeper. The wide range of aquatic plants available includes some of the most beautiful ornamental plants of all. Water plants are so diverse that they encompass both the smallest and the largest leaves in the plant world, from the miniature duckweed (*Wolffia arrhiza*) to the huge Amazonian water lily (*Victoria amazonica*). The A–Z listings of marginals, floaters, oxygenators, deep-water plants and water lilies presented in this chapter represent a core selection to satisfy the varied needs of most pond owners and are widely available.

marginal plants

The shallow water and waterlogged soil at the fringe of the pond are inhabited by specialist aquatics known as marginals, which thrive in various depths of water, from mud up to 15–23cm (6–9in) above their root systems. Different species of marginals have different degrees of depth tolerance, and it is important to check that shallow-water plants are not drowned by water that is too deep. Short or seasonal changes in level will be tolerated in most cases and deciduous marginals will tolerate more water over their crowns in winter than in summer.

Vigorous species of marginals are prevented from invading the centre of the pond only by the water becoming too deep for them to survive, and this is the reason why marginal shelves should be built only 23cm (9in) deep when constructing a pond. Natural ponds without distinct shelves gradually become swamped by vigorous marginals such as the reedmaces (*Typha*) and reeds (*Phragmites*), which form semi-floating rafts of tangled root systems.

Displays of marginal plants in aquatic centres will often look tired at the end of the summer since plants packaged for sale in small containers quickly become pot-bound and deficient in plant nutrients. Late spring is a good time to select and purchase plants, with the new growth thrusting from the fresh compost.

ACORUS CALAMUS (SWEET FLAG)

This is a vigorous species and highly suitable for the wildlife pool. It bears a strong resemblance to the iris and has a most distinctive scent when bruised. It provides good cover for waterfowl but is easier to keep in check than the more common reedmace (*Typha*).

Growing to 60–90cm (2–3ft) in height, the long, bright green, glossy, sword-like leaves have a distinct midrib and part of the leaf edge has quite noticeable crinkling. The unusual pale green flower resembles a small horn, emerging at an angle just below the tip of a leaf. Grow these plants in full sun in water up to 23cm (9in) deep. Propagate them

Left: *Acorus calamus* 'Argenteostriatus' (Variegated sweet flag). Although having a looser form than the variegated flag iris, the variegated sweet flag does retain its variegation all summer.

by dividing the congested rhizomes or cut into pieces 10–15cm (4–6in) long.

ACORUS CALAMUS 'ARGENTEOSTRIATUS'

Similar in habit and growth but not quite so vigorous is the variegated form of the sweet flag, which has a cream-striped variegation on the sword-like leaves that gives added impact.

ACORUS GRAMINEUS (JAPANESE RUSH)

Although in the same genus as the sweet flag, there is little chance of confusing these species since the Japanese rush is very much smaller and has a fan-like growing habit, with glossy, two-ranked, almost sedge-like leaves that are 8–35cm (3–14in) long and semi-evergreen. Although a perennial, it is not entirely hardy in temperate winters and is frequently lost. It will tolerate a very wide range of soil conditions, even to being submerged in small containers in aquaria. When grown in pond margins, plant in full sun in shallow water no deeper than 5–10cm (2–4in). Propagate by cutting up small pieces of root 5–10cm (2–4in) long.

ALISMA PLANTAGO-AQUATICA (WATER PLANTAIN)

A hardy deciduous perennial growing to 76cm (30in) in a sunny position, with rosettes of oval, grey to grey-green, semi-upright ribbed leaves which have long leaf stalks that emerge well above the water. The tiny three-petalled, pinky-white flowers are borne in whorls on a pyramidal spike in midsummer. It produces seed profusely which can be carried some distance on the water surface, germinating in any available soft mud. It is an ideal plant for the wildlife pool, growing in water 15–23cm (6–9in) deep, where it provides a valuable source of food for birds in its seeds. Unlike most aquatic seed, the seed of the water plantain can be stored dry for up to a year and when sown will germinate within 2–5 weeks, producing a fully grown plant within 15 months.

tip

- *Try the highly attractive cultivars of the species* Acorus gramineus, *such as 'Variegatus', which has striped creamy leaves, 'Oborozuki' with yellow leaves and 'Yodo-no-yuki' with pale green variegated leaves.*

Right: Caltha palustris var. *alba* (White marsh marigold). Much more compact than the common yellow marsh marigold, this white form will flower in early spring in mild areas.

BUTOMUS UMBELLATUS (FLOWERING RUSH)

A hardy perennial that is a strong contender for a prime position in the wildlife pool. It has long, pointed, narrow, twisted leaves that are dark green in colour and grow to 1m (3ft) long, with sheathed triangular bases. The elegant flowers are borne above the leaves in a rounded pink flowerhead 10cm (4in) across. It thrives best in rich mud or shallow water no deeper than 8–15cm (3–6in) and in full sun. The plant can be propagated by dividing the rootstock in spring.

CALLA PALUSTRIS (BOG ARUM)

A hardy perennial growing to 25cm (10in), which in mild winters is semi-evergreen, with long, conspicuous creeping surface roots and round to heart-shaped, glossy, mid-green leaves that are firm and leathery. The white flowers appear in spring resembling small, flattened arum lilies, followed by clusters of red or orange berries. Grow in full sun in water no deeper than 5cm (2in). The scrambling surface root can be divided in spring, so long as each cut piece contains a bud.

CALTHA PALUSTRIS (MARSH MARIGOLD, KINGCUP)

The marsh marigold thoroughly justifies its status as one of the most popular marginals for a small pond, normally reaching 15–30cm (6–12in) high with long-stalked lower leaves and stalkless upper leaves. The dark-green leaves are nearly round, heart-shaped at the base and have toothed margins.

The beautiful waxy, yellow, buttercup-like flowers brighten the spring scene, particularly when planted alongside the blue flowers of the grape hyacinth (*Muscari*). They look best in groups in a sunny or part-shaded position at the very edge of the water, tolerating waterlogged or even slightly submerged conditions in winter and a small degree of drying-out during the summer.

The white marsh marigold *C. palustris* var. *alba* varies from the common marsh marigold by having white flowers with yellow centres and a slightly more compact growth. The double marsh marigold *C. palustris* 'Flore Pleno' covers its foliage with a mass of double yellow flowers, often producing a second flush in the summer.

In contrast, the giant marsh marigold *C. palustris* var. *palustris* (sometimes listed as *C. polypetala*) is a massive plant which is only suitable for growing by the side of large ponds, where it can send its large yellow flowers on long stems as high as 1m (3ft). The dark green leaves are as much as 25–30cm (10–12in) across, forming strong hummocks of foliage that can flourish in shallow water up to 10–13cm (4–5in) deep.

Apart from the double form, all the marsh marigolds can be propagated by seed sown when fresh and should germinate in a couple of weeks. Alternatively, the plants can be divided immediately after flowering.

Left: *Cyperus longus* (Sweet galingale). It is vital to keep this containerized in a small pool as it will invade any moist soil near the pool edge. A perfect plant for the larger wildlife pool.

COTULA CORONOPIFOLIA (BACHELOR'S BUTTONS, BRASS BUTTONS, GOLDEN BUTTONS, WATER BUTTONS)

The common names give a clue to the appearance of this bright little marginal for shallow water 8–10cm (3–4in) deep in full sun. It is an annual or short-lived perennial with several creeping succulent stems 15–30cm (6–12in) high and fresh green, strongly scented toothed leaves. The plant is covered with masses of disc-shaped yellow flowers about 1.3cm (½in) in diameter. It tends to die down in the winter but regenerates easily in the spring from the masses of seed produced throughout the year.

CAREX ELATA 'AUREA'

Carex is the collective name for the sedges, which is a very large group of plants mainly enjoying marshy, acid conditions and in many cases shallow water, in sun or partial shade. All have grass-like narrow leaves and triangular flower stalks bearing flowers in brownish spikes.

This hardy species, often listed as 'Bowles' Golden Sedge', is one of the most decorative and particularly valuable in bringing a touch of yellow to the water's edge. Growing to a height of 38cm (15in), it grows in dense tufts of bright, grassy leaves. Brown male flowers appear in spikes 2.5cm (1in) long above stalkless green female flowers on spikes 1.3–4cm (½–1½in) long in early spring before the leaves turn their bright yellow. It is easily propagated by simply dividing the clumps in spring or early summer.

CYPERUS LONGUS (SWEET GALINGALE)

The sweet galingale is one of the few hardy members of the genus and an attractive colonizer of muddy banks in full sun or partial shade. It can spread very quickly so should be kept restricted to containers in small ponds. The stems are almost triangular, growing to 60–120cm (2–4ft), and bear interesting, bright green, stiffly ribbed leaves that radiate from the top of the stem like the

Right: *Houttuynia cordata* 'Chameleon'. A striking plant that is tolerant of a wide range of moisture levels in the soil. Note, however, that in mild areas, it can spread.

ribs of an umbrella. The brown flowers are rather inconspicuous spikelets interspersed amongst the leaves. This is a good plant for the wildlife pool, particularly in late summer and autumn when the arching brown flowers come into their own. It is easily propagated by dividing the roots.

DARMERA PELTATA, FORMERLY *PELTIPHYLLUM PELTATUM* (UMBRELLA PLANT)

Falling between the category of a moisture-lover and a marginal, this waterside plant makes a bold impact at the edge of water in sun or partial shade, where its large round leaves can be reflected. It is a hardy perennial that produces its pink flowers in early spring well before the leaves. The small, starry pink flowers are borne on long, slender, red-tinted stalks reaching 30–60cm (1–2ft) on rounded heads. The leaves appear in spring after the frosts and, like the flowers, are held well above the ground on slender stalks, often over 1m (3ft) high, in sheltered positions. The plant gets its name from the shape of the leaves, which form large, deep green, heavily veined plates that are deeply lobed and coarsely toothed. They reach nearly 60cm (2ft) in diameter and turn beautiful shades of red in the autumn. The roots form thick surface rhizomes, extremely useful for stabilizing the muddy banks of a pond. It should not be grown where water covers the roots and is ideal in muddy, almost saturated soil. To propagate, simply divide the roots in spring.

ERIOPHORUM ANGUSTIFOLIUM (COTTON GRASS)

Cotton grass is grown for its very conspicuous white tassels of cotton-like flowers that stand out dramatically in dull conditions, although it enjoys full sun. Out of flower it is a rather dull plant with short, leafy, angled stems usually no more than 30cm (12in) high. It should not be covered by more than 5cm (2in) of water for any length of time. It is easily increased by dividing the clumps.

GLYCERIA MAXIMA VAR. *VARIEGATA* (MANNA GRASS, SWEET GRASS)

Glyceria is a highly ornamental, hardy aquatic grass which deserves a place in the ornamental pond provided it is kept in check in an aquatic container. It is extremely invasive if let loose in an unrestricted root run. It has very striking leaves which are striped cream, white and green and in the spring the young leaves are flushed with pink. It is very easy to grow, reaching a height of 60cm (2ft) in water up to 15cm (6in) deep in full sun. The flowers form greenish spikelets in summer. It is very easily increased by dividing the roots.

HOUTTUYNIA CORDATA

This is a very useful clump-forming hardy perennial, growing 15–30cm (6–12in) high with spreading roots and erect, leafy, red stems. The bluish-green, leathery, pointed leaves give off a pungent smell when crushed. In spring, spikes of rather insignificant yellowish-green flowers are produced, surrounded by white bracts. It is on the borderline of hardiness and benefits from a thick leafy mulch in autumn. It can be invasive and is best grown in containers in a small pool. It will not tolerate water deeper than 2.5–5cm (1–2in) and prefers a semi-shaded position. *H. cordata*

Left: *Iris pseudacorus* 'Variegatus'. One of the most striking of variegated leaves for the pool margins that unfortunately fades to a dull green as the summer progresses.

'Chameleon' is a highly colourful plant with leaves that are splashed with crimson, green and cream. This cultivar must have full sun to show off its complete colour range. Propagation is easy by dividing the extensive matted rhizomes in the spring.

tip

- *Bog bean is the ideal choice of marginal for quickly colonizing the shallow water of a large sunny pool, with its thick spongy rootstocks rapidly spreading just underneath the surface.*

IRIS

It is hard to imagine any pond margin without the sword-like leaves of iris thrusting from the water surface. The following two species are the most common and popular. They are both easily increased by dividing the thick roots in spring or early summer.

IRIS LAEVIGATA

One of the finest irises for shallow water (10cm/4in deep) in sun or partial shade, producing clumps of sword-shaped, soft green leaves without a midrib growing to 0.6–1m (2–3ft). The sparsely branched stems produce two to four broad-petalled, beardless blue flowers in early summer. When not in flower, the leaves may be confused with those of the Japanese iris *I. ensata* (formerly *I. kaempferi*), which has large showy flowers but is not a true marginal since it dies if the roots remain under the water in winter. It is distinguished from *I.*

laevigata by its distinct midrib. There are many excellent cultivars of *I. laevigata* but one of the most impressive is *I. laevigata* 'Variegata' which has pale, lavender-blue flowers and lovely cream-and-white-striped leaves, making it one of the most sought-after marginal aquatics.

IRIS PSEUDACORUS (YELLOW OR FLAG IRIS)

A vigorous perennial with strong, stiff, bluish-green, sword-like leaves growing to 1m (3ft), emerging from a thick rhizome that binds the soil surface. The tall, branched flower stems bear as many as ten beardless yellow flowers on each stem. Each flower has radiating brownish veins and a deeper orange spot in the throat. An ideal species for the margins of the wildlife pool, it grows in 15–30cm (6–12in) of water in sunny or shaded positions. *I. pseudacorus* 'Variegata' is suitable for small ponds if containerized and has striking creamy-striped variegation which appears in spring and gradually fades as the summer advances.

JUNCUS ENSIFOLIUS

This small member of the rush family is a charming hardy marginal for the side of a small pool, where it grows to no more than 30cm (12in) high in sun or partial shade. Rushes mainly occur in marshes and bogs with a few species growing in shallow water. They prefer very damp, mostly wet banks where they look best in groups. They have a habit and general appearance similar to grasses, often with flattened leaves sheathed at the base. The majority are not included in ornamental pond margins, but this little species is one of the exceptions. It is particularly good for the side of a stream, producing neat tufts of mid-green grassy foliage and attractive round brown flower spikes. It will not tolerate water deeper than 5cm (2in). It is simple to increase by dividing the grassy clumps in the spring and summer.

LYSICHITON (SKUNK CABBAGE)

The plant's common name refers to the unfortunate scent of its flowers, but in every other way it is an attractive and intriguing hardy perennial marginal which never fails to attract attention in early spring, when its arum lily-like flowers appear before the leaves in wet soil. The unpleasant smell is hardly noticeable with the flowers held so close to the ground in the cool spring weather. The white skunk cabbage *L. camtschatcensis* has white flowers and is less vigorous than the more common yellow-flowered *L. americanus*, making it more suitable for the margins of small ponds. It seldom reaches more than 60cm (2ft). The leaves are mottled and paddle-shaped, growing semi-upright from ground level with little or no leaf stalk. It is on the borderline of tolerating shallow water over the roots, but so long as there is a deep saturated soil it will not mind short periods of being submerged in shallow water. Grow in full sun and propagate by seed.

MENYANTHES TRIFOLIATA (BOG BEAN, BUCKBEAN, MARSH TREFOIL)

Bog bean is a hardy perennial marginal that has attractive foliage of clover-like, shiny, olive-green leaves made up of three leaflets with a long leaf stalk that clasps the spreading roots with a broad sheath. The flower is particularly appealing, forming dense spikes 25–40cm (10–16in) above the water surface of dainty, frilled, white to purplish flowers that emerge from pink buds. It should be containerized in a small pool and any escaping roots cut back hard each spring. In the larger wildlife pool it provides excellent cover for submerged creatures in the shallow margins and can spread to form clumps of 1m (3ft) or more in one year. The spreading roots can easily be divided if more plants are required.

maintains vigour and increases stock. 'Mermaid' is an improved cultivar that is more free-flowering than the species.

PONTEDERIA CORDATA (PICKEREL WEED)

This hardy perennial marginal is undoubtedly one of the most decorative blue-flowered aquatic plants available. It forms a robust, tidy plant growing to 45–60cm (18–24in) high. The thick, creeping rootstock supports shiny, erect leaves that are heart-shaped and olive green with exquisite swirled markings. A soft blue flower spike appears from a leaf bract at the top of the stem. It will succeed with up to 13cm (5in) of water above its crown, and will show off its full flowering potential when baked in full sun. It can be divided once the plant is actively growing in late spring.

MYOSOTIS SCORPIOIDES (WATER FORGET-ME-NOT)

This is one of the most sought-after hardy marginals for ponds of all sizes in full sun. It has a slightly looser and more delicate habit of growth than most of the more rampant marginals, and delightful light blue flowers with yellow centres in midsummer. The angular stems are almost upright, becoming erect at the tips, and reach a height of 23–30cm (9–12in). Cover with no more than 8cm (3in) of water and regularly divide established clumps in spring, which both

RANUNCULUS FLAMMULA (LESSER SPEARWORT)

This is a hardy perennial member of the large buttercup family and is suitable for small pools, unlike many of the aquatic ranunculus which can be rank growers. Although most at home in the shallows of a wildlife pool in full sun or part shade, it can easily be containerized and creates a low-growing (60cm/2ft high) spread of yellow flowers for a long season in the smaller pond. It has semi-prostrate, reddish

Left: The delicate, semi-upright growth of *Myosotis scorpioides* (Water forget-me-not) makes a welcome change in form.
Right: *Pontederia cordata* (Pickerel weed) is one of the few blue flowers for the waterside in late summer.

stems and dark green lance-shaped leaves 1.3–3cm (½–1in) long. The bright yellow cup-shaped flowers are 2cm (¾in) across and borne in clusters in early summer. It is easily propagated by dividing clumps in spring or early summer.

SAGITTARIA (ARROWHEAD)

The arrowheads are reliable hardy perennial marginals suitable for container growing or for the freedom of the larger wildlife pool. The plant produces fine, shiny, arrow-shaped leaves up to 46cm (18in) long and white flowers that are borne on triangular stems in whorls of three. *S. sagittifolia* 'Flore Pleno' is a very handsome double form, sometimes sold as *S. japonica* (Japanese arrowhead), with magnificent round, double white flowers about 2.5cm (1in) across arranged around a spike.

The arrowheads produce overwintering walnut-sized tubers at the ends of the roots, and these become detached in the autumn. The tubers are rich in starch and attractive to wildfowl, giving the plant yet another common name of 'duck potato'. It prefers full sun and should be planted in shallow water no deeper than 5cm (2in), since flowering will be restricted in deeper water. It is propagated by dividing the roots in the spring or summer.

SCHOENOPLECTUS LACUSTRIS SUBSP. *TABERNAEMONTANI* (BULRUSH)

Although the common name of bulrush is frequently associated with the large rank plants of *Typha*, which bear brown poker-like flowers in the autumn, this plant is the true bulrush, and may be listed in some catalogues under its former name of *Scirpus*. Bulrushes are found naturally in marshes and shallow water, and are identified by their very long thin stems growing from rampant roots which require keeping in check.

There are two attractive variegated forms, which are very effective when grown in groups around the edge of a large pond in water 23–30cm (9–12in) deep in full sun. 'Zebrinus' or zebra rush is perhaps the most intriguing as the hollow cylindrical stems have distinctive markings for an aquatic. As its name suggests, it has horizontal stripes like porcupine quills on its narrow leaves and grows to 1–1.2m (3–4ft). The other variegated form, 'Albescens', has longitudinally striped leaves and tends to be more vigorous, growing to 1–2m (3–6ft). The flowers are inconspicuous, appearing at the tips of the long leafless stems in the form of a brown spikelet. The bulrushes are easily propagated by dividing the extensive roots.

TYPHA MINIMA (DWARF REEDMACE)

This is the only reedmace suitable for a very small pond, growing to only 30–46cm (12–18in). All the other typhas are much too large and their roots can damage flexible liners. The leaves are needle-like and the dark-brown flower spikes are round, unlike the pokers of the more vigorous species. It readily takes to container growing and will tolerate a water depth of 10–15cm (4–6in) in full sun. Propagate by dividing the roots in the spring.

Left: An attractive oxygenator, milfoil (*Myriophyllum aquaticum*), enjoys creeping out of the water.

floating plants

This selection features the most commonly available floating plants. Floaters are extremely valuable in the initial stages of the establishment of a new pond, providing much-needed shade.

AZOLLA FILICULOIDES (FAIRY MOSS)

A small perennial aquatic fern, growing to only 2mm (⅛in high), that forms clusters of soft, pale green leaves which turn purplish-red in the autumn. Each leaf is attached to a single fine root. It can be invasive, so thin regularly. It survives the cold winters by producing overwintering bodies that sink to the bottom, resurfacing when the water warms up in late spring. Grow in full sun. Since it spreads so rapidly, propagate by simply removing offshoots at any time.

EICHHORNIA CRASSIPES (WATER HYACINTH)

This tender plant will not survive the winter outdoors in temperate climates. It becomes commercially available in summer and provides a striking sight on the pool surface for a few months in the hot weather. Growing to 46cm (18in) high and wide, it has rosettes of shiny, pale green, 15cm (16in) wide leaves with swollen spongy bases, which act as floats. In very warm summers it produces pale blue hyacinth-like flowers up to 15–23cm (6–9in) high. Long, feathery, dangling roots, purplish-black in colour, provide the perfect medium for spawning goldfish. It spreads like a strawberry by fast-growing stolons, which in warm climates have caused the plant to be prohibited because it chokes waterways. No such problem exists in temperate climates, and it will only be successful in warm, sunny areas. Overwinter in a frost-free greenhouse on trays of moist soil. Propagate by removing offshoots at any time.

STRATIOTES ALOIDES (WATER SOLDIER)

Water soldiers are hardy, semi-evergreen perennials that can form quite extensive stands of prickly rosettes of olive-green leaves, 50cm (20in) long and 2.5cm (1in) wide, with serrated edges. The tips of the leaves frequently emerge above the water surface. During the summer the whole plant tries to surface in order to produce cup-shaped white, sometimes pink-tinged flowers about 4cm (1½in) across. It readily produces offsets that are easily detached for propagation purposes.

tips

- *The intertwined mass of stems of the starwort provides homes for abundant minute pond life, and the young leaves are a particular delicacy for goldfish.*
- *Golden club can also be grown as a marginal in shallower water, where the leaves grow to a larger size and their undersides are exposed.*

oxygenators

Oygenators are available throughout the summer, mainly in the form of bunches of unrooted stems. These plants are all easily propagated by taking softwood cuttings 20–30in (8–12in) long from the new growth during the summer.

CALLITRICHE (STARWORT)

The starworts are small slender plants that generally grow in a tight mass and occur in a wide variety of habitats, mainly in temperate areas. They are characterized by their terminal rosettes of light-green leaves 5–6cm (2–2½in) across which, when floating on the water surface, give rise to their common name. *C. stagnalis* is a hardy species with slender branching stems that root at the nodes. The leaves are nearly circular, forming rosettes at the stem tips.

CERATOPHYLLUM DEMERSUM (HORNWORT)

These plants usually form clusters of leaves 60–90cm (2–3ft) long with whorls of stiff, slender, dark green leaves crowding towards the apex. They are particularly useful in more shaded ponds where there would not be enough light for most other submerged oxygenators. Hornworts are frequently found free-floating or very loosely anchored in the bottom mud in both still and moving water. The brittle whorled leaves are dark green, 1.5–4cm (½–1½in) long and forked. Like most oxygenators, it has inconspicuous white male and green female flowers borne in the leaf axils. In order to overwinter, the tips of the shoots shorten and thicken, then break off and sink to the pool bottom.

MYRIOPHYLLUM (MILFOIL, PARROT FEATHER)

The milfoils show a great diversity of habit, and many love to creep out of the water and scramble onto the surrounds of the pool. *M. aquaticium* is slightly tender and must be well submerged in temperate climates to survive the winter. It is, however, extensively used in outdoor pools where the graceful foliage and stems take root in the wet soil above the water line. The stems can grow 50–150cm (20–60in) long with leaves 2.5–5cm (1–2in) long in whorls of 4–6, each divided into 4–8 bright green segments.

POTAMOGETON CRISPUS (CURLED PONDWEED)

The curled pondweed is one of the few of its species that is suitable for decorative ponds, since most quickly choke out other submerged plants. The stems are capable of growing to a staggering 4m (13ft) or more, bearing narrow stalkless leaves about 8cm (3in) long and 0.5–1cm (⅕–⅜in) wide. The leaves are most attractive, being almost translucent and wavy-edged, resembling seaweed, varying from green to reddish brown. It tolerates cloudy water better than any other oxygenator.

Below: *Callitriche stagnalis*. One of the most efficient oxygenators which continues functioning in the winter months rather than dying off like some of the more tender submerged plants.

Left: Aponogeton distachyos (Water hawthorn). The superb purplish-brown anthers need to be seen at close quarters to be fully appreciated. The leaves of these deep-water plants can form a strong contrast to the rounded ones of water lilies.

deep-water plants

Although the water lilies dominate this group of aquatics, there are a few species of deep-water plant that add variety to the circular pads of water lilies and provide some interesting flowers. As well as being decorative, their leaves also provide shade.

APONOGETON DISTACHYOS (WATER HAWTHORN, CAPE PONDWEED)

A frost-hardy perennial aquatic with oblong, bright green leaves up to 20cm (8in) long by 8cm (3in) wide that can become almost evergreen in mild winters. The strongly scented flowers are often produced in two flushes, the main flush in spring and a second most welcome surprise in the autumn. The beautiful white flowers with purple-brown anthers are 10cm (4in) long and held above the water. A very tolerant plant, particularly of shade, it extends the flowering season and its long leaves add interest to the water surface. Grow in water 30–90cm (1–3ft) deep, and propagate by dividing the rootstock in spring into 5cm (2in) lengths containing 2–3 buds.

ORONTIUM AQUATICUM (GOLDEN CLUB)

A slow-growing, hardy perennial for water 38–46cm (15–18in) deep, with handsome large, bluish-green, velvety lance-shaped leaves with a silvery sheen on the undersides. In water deeper than 30cm (12in) the leaves float on the surface, growing 13–30cm (5–12in) long and 13cm (5in) wide with a waxy covering. The unusual white pencil-like flowers emerge from the water surface tipped with yellow, resembling small golden pokers. Propagation is by seed sown when fresh in midsummer.

POLYGONUM AMPHIBIUM (WILLOW GRASS, AMPHIBIOUS BISTORT)

A hardy perennial aquatic which, as its name suggests, will grow in terrestrial as well as aquatic conditions. It is at its best in a sunny position in water, where the lovely dense spikes of pink flowers 5cm (2in) long thrust above the surface in midsummer. It has long-stalked, floating leaves 8–10cm (3–4in) long and 2–4cm (¾–1½in) wide carried on stems 30–90cm (12–35in) long which root from the stems. Growing in water up to 46cm (18in) deep, it makes an attractive plant for the wildlife pool or for pools with fluctuating water levels. It has the bonus of some lovely tints in the leaves in autumn. Propagation is by dividing pieces of the stem, each with a portion of root, as they spread across the bottom mud.

Right: *Nymphaea* 'Pink Sensation'. Extremely free flowering with the bonus that the flowers stay open longer into the afternoon than most other cultivars.

water lilies

All the water lilies bear the generic name of *Nymphaea*, and the water lily family is one of the oldest families of water and marsh plants scattered around the world. This selection features the hardy varieties that can stand the pond freezing over in the winter, but in warmer climates there is an equally comprehensive range of tropical types that bloom either at night or during the day. The flowers vary in size but are generally always in proportion to the leaf, ranging from 2.5cm (1in) in the pygmy varieties to 30cm (12in) in some of the tropicals. There are water lilies suitable for every size of pond, from barrels and tubs to extensive pools. All should be given a sunny, sheltered position without any water turbulence on their leaves. Propagate by dividing the roots in summer.

The planting depth specified refers to the depth of water above the growing point, not the depth of pond. The spread refers to the average area that the leaves will eventually cover, although in small planting containers they may not achieve these sizes.

PINK

N. 'Darwin' (formerly 'Hollandia')

A double, peony-style flower 15–19cm (6–7½in) across with light pink inner petals and lighter pink, almost white outer petals surrounding pinkish-yellow stamens. The new leaves are brownish turning green, round in shape and 25–28cm (10–11in) across. Spread 1.2–1.5m (4–5ft); planting depth 30–60cm (12–24in).

N. 'Perry's Pink'

Star-shaped flowers 15–18cm (6–7in) across with rich pink petals and yellow to orange stamens. The new leaves are reddish purple turning green and round in shape, up to 28cm (11in) across. There is an unusual red dot in the centre of the flower. For best flowering it should be planted in a large container. Spread 1.2–1.5m (4–5ft); planting depth 30–60cm (12–24in).

N. 'Pink Sensation'

Cup-shaped flowers becoming star-shaped, 13–15cm (5–6in) across with pink petals surrounding yellow and pink stamens. Purplish young leaves mature to round green leaves up to 25cm (10in) in diameter. One

Bronzed red young leaves mature to round green leaves 15cm (6in) in diameter. A very good water lily for cooler situations and barrels or small pools. Spread 90cm (3ft); planting depth 15–30cm (6–12in).

N. 'James Brydon'
Cup-shaped flowers are 10–13cm (4–5in) in diameter with brilliant rose-red petals and orange-red stamens. Purplish-brown young leaves blotched with dark purple mature to round, green leaves 18cm (7in) in diameter. A very popular variety for barrels or medium-sized pools for its shape of flower and free-flowering habit. Spread 0.9–1.2m (3–4ft); planting depth 30–46cm (12–18in).

N. 'Lucida'
Star-shaped flowers are 13–15cm (5–6in) across with red inner petals and pink-veined, whitish-pink outer petals surrounding yellow stamens. The mature oval leaves grow to 25cm (10in) long and 23cm (9in) wide, with large purple blotches. Suitable for any sized pool, it is free-flowering and has particularly attractive leaves. Spread 1.2–1.5m (4–5ft); planting depth 30–46cm (12–18in).

of the best pinks whose flowers stay open late into the afternoon. Spread 1.2m (4ft); planting depth 30–46cm (12–18in).

RED

N. 'Charles de Meurville'
Star-shaped flowers 15–18cm (6–7in) in diameter with dark, pinkish red inner petals and pink outer petals surrounding brilliant orange stamens. The very long leaves are dark green with light green veins, and 25cm (10in) long by 20cm (8in) wide. It is one of the first water lilies to flower with almost plum-coloured blooms occasionally streaked with white. Spread 1.2–1.5m (4–5ft); planting depth 46–60cm (18–24in).

N. 'Escarboucle'
Cup-shaped flowers that become star-shaped, about 15–18cm (6–7in) across with bright vermilion-red petals, the outer ones tipped white, with deep orange stamens. The brown-tinged young leaves mature to round green leaves 25–28cm (10–11in) in diameter. One of the best reds for medium and large pools, staying open later in the afternoon than most other red varieties. Spread 1.2–1.5 (4–5ft); planting depth 30–60cm (1–2ft).

N. 'Froebelii'
Cup-shaped flowers that become star-shaped, 10–13cm (4–5in) across with burgundy-red petals and orange-red stamens.

N. 'Radiant Red'
A star-shaped flower with long sepals, 13–15cm (5–6in) across with slightly flecked deep red petals and orange stamens. The new leaves are slightly bronzed before turning green and almost round in shape, up to 25cm (10in) across. Spread 1–1.2m (3–4ft); planting depth 30–46cm (12–18in).

N. 'Vesuve'

Star-shaped fragrant flowers are 18cm (7in) in diameter with inward-pointing, glowing red petals that deepen in colour with age surrounding orange stamens. The almost circular green leaves are 23–25cm (9–10in) across. It has a long blooming season in addition to opening early in the morning and closing late in the afternoon. Spread 1.2m (4ft); planting depth 30–46cm (12–18in).

N. 'William Falconer'

Cup-like flowers are 10–13cm (4–5in) in diameter with very deep red petals and burgundy-red stamens. The new leaves are purple, maturing to almost round green leaves 20cm (8in) in diameter. This is a good variety for cool areas since it resents too much heat and will stop flowering in very hot spells. Spread 1m (3ft); planting depth 30–46cm (12–18in).

WHITE

N. 'Gladstoneana'

A star-shaped flower 13–18cm (5–7in) in diameter with white petals surrounding yellow stamens. Bronzed young leaves mature to almost round, wavy-edged green leaves 28–30cm (11–12in) across with crimped margins along the lobes. It is very free flowering and best suited to larger pools. Spread 1.5–2.4m (5–8ft); planting depth 46–60cm (18–24in).

N. odorata minor

The small star-shaped flowers are 8cm (3in) in diameter with white petals surrounding prominent golden-yellow stamens. The green leaves are almost round and 8–10cm (3–4in) in diameter with dark red undersides. It is most successful in tubs or shallow pools but must have full sun. Spread 60cm (2ft); planting depth 23–30cm (9–12in).

N. 'Virginia'

Star-shaped, almost double fragrant flowers 15–18cm (6–7in) in diameter with creamy, pale yellow petals in the centre and pure white petals on the outside surrounding yellow stamens. The young leaves are green with heavy purple blotching, which becomes restricted to the perimeter of the mature egg-shaped leaves 25cm (10in) long and 23cm (9in) wide. A free-flowering classic water lily. Spread 1.5–1.8m (5–6ft); planting depth 30–60cm (12–24in).

Left: *Nymphaea Nymphaea* 'Vesuve'. A well tried older red cultivar with a long flowering season.
Above: *Nymphaea* 'Marliacea Chromatella'. One of the finest and most reliable yellow water lilies that has beautifully marbled leaves when young.

YELLOW

N. 'Marliacea Chromatella'

Cup- to star-shaped flowers are 15cm (6in) in diameter with broad, incurved, canary-yellow petals and golden stamens. Coppery young leaves with purple streaks mature to attractive purple-mottled, round green leaves 15–20cm (6–8in) in diameter. One of the best long-standing, reliable yellow water lilies which will perform well in any sized pool. Spread 1.2–1.5m (4–5ft); planting depth 30–46cm (12–18in).

N. tetragona 'Helvola'

Cup-shaped, later star-shaped flowers no more than 5–8cm (2–3in) in diameter with yellow petals and yellow stamens. The leaves are oval, 13cm (5in) long and 9cm (3½in) wide, heavily mottled and purple-blotched with purple undersides. This delightful small water lily makes a perfect plant for a barrel or sink. Spread 60cm (2ft); planting depth 15–23cm (6–9in).

8 TROUBLE-SHOOTING

The difference between a pond and any other ornamental garden feature is that it forms a miniature world of tiny micro-organisms that work together to create a balanced living entity. With good construction and planting techniques, this balance can be achieved relatively quickly and sustained indefinitely with limited assistance. But from time to time problems occur which are not directly related to the pests and diseases that can attack plants and fish. The following pages highlight the most frequently encountered pond problems and offer practical strategies for solving them effectively.

discoloured water

This is by far the most common problem of small garden ponds. If left unsolved, it can spoil much of the pleasure of the pond, eventually leading to a total loss of interest.

GREEN WATER

This is the prime discoloured water problem and it relates to the interrelationship of size, depth and type of planting. Green water is caused by the growth of tiny single-celled algae that flourish in warm water receiving full light and charged with mineral salts. This pond condition is most likely to be encountered shortly after a new pond has been filled, since the water will probably be minerals-rich mains water and the planting in the pond will not have become established. Clear water can turn to green water almost overnight, but the reversal can be just as dramatic and swift when sufficient factors all act together at the same time to destroy the algae.

If the pond is relatively new and the correct mix of plants has been introduced, it is best to leave it alone entirely for a few weeks, sometimes even months, until the right balance of growth is achieved. Refilling the pond, tempting though this might be, is not the answer. It will green up in just the same way shortly after the refill. If the condition persists in an established pond, there is likely to be one or possibly more causal factors.

Planting remedies

Among the causes of green water may be a lack of submerged planting to starve out the green algae. In this case it is a relatively easy matter to remedy the deficiency by planting more. Ensure that the guideline of 5 bunches to every 1sq m (10 sq ft) is followed even in an established pond where green water persists. If the surface of the pool is almost clear, provide shade for the water by adding some floating

Left: Without the circulation from the fountain and no planting, the pool water quickly becomes cloudy as algae take hold.
Below left: In contrast, the geyser jets introduce plenty of oxygen into this pool. Supplementary planting ensures that the pool is even less likely to turn green.

plants so that at least two-thirds of the surface is covered. Ideally, the leaves of water lilies should achieve this coverage, but in the spring and in newly planted ponds this is not easily achieved.

With the additional plants now in place, the clear water will still be frustratingly slow to clear, so a great deal of patience is required. If the planting fails to solve the problem in the long term, there may be a basic problem in the shape of the pond. Small shallow, saucer-shaped ponds are the most vulnerable to greening since they heat up very quickly and the light level is very high at all depths. These shallow ponds lose a higher proportion of their volume to evaporation than deeper, larger ponds, and if water loss is compensated by adding tap water too often, a fresh supply of mineral salts will be introduced for the algae each time.

Chemical treatments

The ratio of surface area to volume in small shallow ponds is probably too high to establish the right balance and chemicals may have to be used to achieve and maintain water clarity. There are a great variety of these on the market and they work by killing the algae. One application of algaecide, however, will not be a permanent answer and it will not be long before another application is required as the algae start to grow again. In order to apply algaecides at the correct dosage, you will need to calculate the volume of your pond. Without this information, there is a very great risk of over-applying the chemical, causing long-term damage to both fish and plants.

First calculate the surface area of the pond by multiplying the length by the width for rectangular or square ponds, working in metres or feet. For kidney- or irregular-shaped ponds, take the maximum length and the maximum width by drawing an imaginary rectangle around the shape. For more complex shapes, divide the pond into two or three sections by drawing imaginary boxes around them, then calculate the area of each box. Add the results together for the total surface area. For circular ponds, apply the following formula involving a mathematical constant called *pi* (3.142), which represents the ratio of the circumference to the diameter of the circle. The surface area of a circular pond is calculated by multiplying the radius of the pond by itself (squaring the radius), then multiplying this figure by *pi* (3.142). The volume of the pond is then calculated by multiplying the surface area by the depth of the pond. If working in metric, multiply this figure by 1,000 to find out the volume in litres, or if working in imperial, multiply by 6.25 for the volume in gallons.

Preventing an influx of nutrients

One final approach in seeking a cure for persistent green water is to ensure that the surrounding contours do not cause surface water to seep into the pool whenever there is heavy rain. Surface water from the garden is likely to be enriched with nutrients, particularly if the lawn surrounding the pond receives regular top-dressings of fertilizer.

BROWN WATER

This is more likely to be a problem in ponds containing large fish such as koi which stir up mud from the bottom. In wildlife ponds where the pond bottom has a thick layer of mud, it can become an almost permanent condition. It does absolutely no harm to any fish present but is more of a nuisance in preventing the bottom from being seen clearly. Algaecides have no effect on this condition and chemicals known as flocculating agents, which cause the fine particles in suspension to clump together and sink, can be used if the disturbance of the bottom is not likely to be repeated. If being able to see big fish on the pool bottom is a priority, a new pond design incorporating filtration should be considered.

BLACK WATER

Many water gardeners, particularly in North America, add a vegetable dye to the water to turn it inky black. This is done to increase its reflective quality and to create interesting visual effects. This practice is not to be confused with a condition that may occur in a pond where too much organic matter, such as tree leaves, has been allowed to settle on the bottom. This excessive organic matter decays and causes the water to become quite black – a serious toxic condition for fish. This is remedied by gradually changing the pond water through two or three partial water changes, preferably with a dechlorinator added to the fresh water. To prevent the same condition recurring the following spring, place a net over the water in the autumn to prevent leaves from entering, and regularly cut back oxygenators.

MILKY WATER

If there is a significant decomposition of dead fish or other creatures that might have died in the water, such as frogs, a slight milky appearance can occur. A complete water change is necessary here, checking at the same time that there is no offending dead fleshy material hidden on the pool bottom.

OILY WATER SURFACE

Decomposing water lily leaves are the most frequent cause of a thin, streaky film on parts of the pond surface. Keep removing the dying water lily leaves before they decompose, and remove the thin, oily patches with newspaper laid on the water surface. Drag the paper off to the side after a minute or two when it will have absorbed the oily material. The use of newspaper is also recommended where a pond is near to heavy conifers that are constantly shedding pollen, dust and fine leaf particles. The paper is excellent at absorbing material that is too light to sink through the very thin skin that water possesses at its surface.

ACID OR ALKALINE WATER

Extremes of water chemistry can occur in certain areas and conditions, although this may not be apparent through discoloration of the water. Very alkaline or hard water is most likely to occur where the mains water is known to be hard and is used to top up a small pond on a regular basis.

Knowing the level of acidity or alkalinity of the pond water is useful to help avoid swinging between extremes, because this can cause problems for fish and with the absorption of nutrients by the plants. Simple water-testing kits are readily available and inexpensive, and should be used once a year to monitor pond chemistry.

If the pond accumulates large quantities of decaying organic matter, it will make the pool quite acid and both fish and plants will be weakened as a consequence. A water change is the recommended solution, and if the mains water is already very alkaline, use rainwater that has been collected in a rainwater barrel. Do not attempt to cure acidity in a pond by adding lime. The change that would result would be too rapid and could cause considerable stress or death to the fish.

Left: A shallow pond or beach makes an ideal environment for blanketweed. The cobbles make it difficult to pull the blanketweed out and chemical control may become necessary.

blanketweed

Although blanketweed is a form of algae, it is not the same type of algae that causes the pond water to cloud and go a pea-green soup colour. Blanketweed is in fact a problem most likely to occur in otherwise crystal-clear water. The name is descriptive of its appearance – it forms large mats of green fibrous matter that consolidate into clumps at all levels of the water. It is an absolute menace in blocking up pump strainers and trapping small fish and tiny pond creatures in its tissue. Blanketweed is at its worst in shallow, alkaline water receiving full sun. Small shallow streams and pebble ponds are particularly prone where it clings to any available extra surface area.

Removal by hand is possible if done on a regular basis, and because of its clinging nature it can be wrapped round a forked stick immersed in the water. There is a huge variety of chemical remedies, from straight herbicides to water conditioners, which tend to reduce the alkalinity of the water, and a small fortune can be spent if you have a large pond with this problem. The stronger the weedkiller, the more likely it is to harm other submerged plants such as water lilies, so you must be extremely careful in dosage rates and regularity of use.

CONTROL STRATEGY

There is no easy remedy for blanketweed, and the following combination of methods should be used in its control.

- Provide adequate shade in the form of water lily leaves.
- Keep surfaces as clean as possible to prevent blanketweed gaining a foothold on awkward shapes.
- Reduce the area of shallow beach edges where the water warms up quickly and replace with deeper vertical sides which are cooler.
- When topping up, use rainwater if possible in areas known to have hard water.
- Keep a convenient forked stick handy so that weed can be removed before large clumps build up.
- Use chemical remedies as a last resort.

Right: Containerized plants should be removed and divided regularly. Plant the young division into fresh compost before returning it to the pond.

neglected overgrown ponds

Most ponds reach the stage where they become so overgrown that a complete clean-out is the only way to remove the accumulation of decomposed debris on the bottom and restart the old established clumps of water lilies. Water lilies can quickly escape from their planting containers and grow in the mud on the pool bottom. Very soon the pond becomes dominated by the foliage, spoiling its appearance and limiting the area in a small pool that can be enjoyed by the fish. When the water lily leaves thrust above the surface, this is indicative of the plant needing to be divided and is best done when the pond itself is cleaned out.

A balance should be kept between the clinically clean pond, which offers no hospitality to submerged or amphibious creatures, and a pool that has become choked with vegetation and full of mud. In the pond-clearing process, a small quantity of both mud and the old pond water should be replaced in the cleaned-out pond so that the microscopic life retains a foothold. If rainwater is available to supplement the refill, this also helps to re-establish the biological balance more quickly.

Left: Water lilies that have been neglected push their leaves well above the surface and produce very few flowers. The only solution is to clean the pond out completely, thin the water lilies and propagate new ones from the young tips.

WHEN TO CLEAR

Mid- to late summer is a good time to organize this, since least damage will be done to the pool population and it will have time to settle down again before the winter. If the pond is emptied too early in the year, hibernating or breeding amphibians will be disturbed. If undertaken in the autumn as the plants and animals become dormant, many of the divided or disturbed plants will rot, and many amphibians could die. Clearing out ponds is a smelly, messy job and best done on a cool overcast day as strong sunshine would cause extra stress to the pool occupants when dislodged to their temporary home.

SAFEGUARDING FISH

Plans should be made to ensure that any fish are not starved of oxygen. A large precious fish would soon die if kept only for a few hours in a temporary container in warm water in which the oxygen supply had been depleted. Large fish will suffer less oxygen starvation if a small aerating pump such as those used in aquaria could be temporarily installed in the holding pool. Use the water from the pond to fill the temporary housing quarters for the fish.

POND-CLEARING PROCEDURE

If there is a submersible pump already *in situ*, use this to connect to a hosepipe and pump out the water to a nearby drain. Without a pump, siphoning may

tips

- *Use special pond gloves for pond-clearing, which have long PVC sleeves welded to thick rubber gloves and elasticated tops to keep your hands and arms totally dry and clean as you work.*
- *If wading into a pond with a flexible liner, avoid wearing boots or waders with studded soles, which might cause damage.*
- *When moving fish in and out of the pond, if they must be held use a wet towel to minimize damage to their scales.*
- *If a fountain spray can be turned on, this will help the fish recover from any oxygen starvation resulting from being in their temporary quarters.*

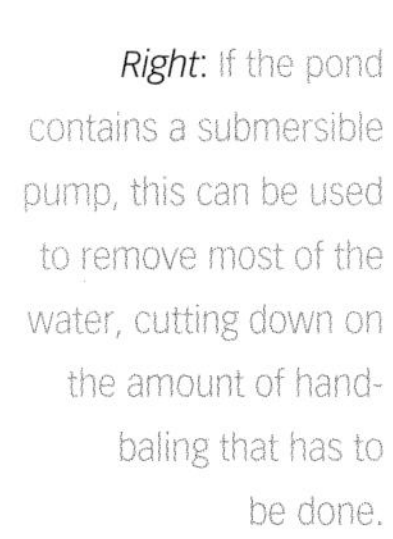

Right: If the pond contains a submersible pump, this can be used to remove most of the water, cutting down on the amount of hand-baling that has to be done.

be quicker than baling out by hand. If fish are present, remove just enough water to make catching them easier. Once the bottom mud is disturbed and the fish become agitated and harder to catch, the job of netting them will have to wait until there is only a fraction of the water left in the bottom of the pond because until that point they become impossible to see.

Remove the marginal plants first and transfer them to a shady spot nearby. If the job lasts no more than a day or two, they will have sufficient water reserves in the wet compost. Having caught as many fish as possible while the water is still clear, wade into the pool and remove the containers of deep-water plants such as water lilies. This is a strenuous job, since water lilies tend to cling to the bottom, so enlist help with lifting them. Cover the plants with wet newspapers and keep in as shady a position as possible. Like the marginals, the water lilies will survive out of water for a day or two as long as the newspaper covering is kept wet. Lift out any containers of oxygenators and give them a hard pruning. These plants should then be submerged quickly in temporary containers, otherwise they will very soon shrivel up and die.

After the vegetation is removed, trawl the muddy water with a wide soft net to catch the remaining fish, then continue to pump out as much water as the pump can cope with. There will be some water remaining which the pump cannot effectively handle, and this should be baled out by hand into buckets at the pond side. This mud can be used on any nearby borders, rather than poured down drains where itr may clog silt traps. Remove the thicker sludge with a rubber bucket or dustpan, taking care in a lined pond not to score the surface of the liner. Keep a small amount of sludge to return to the freshly cleaned pond.

With all the vegetation and fish removed, scrub the sides and bottom of the pond to remove algae. Any dirty water generated from this process can be baled out.

REFILLING

Partially refill the pool to a height that will allow repositioning of the freshly divided water lilies and marginals. Reintroduce oxygenating plants that have been cut back into the shallow water with the small quantity of mud kept from the pond before cleaning.

With the deep-water plants installed, reintroduce the fish from their temporary quarters. Pour back the original pond water used for housing the fish into the pond. Finally, reintroduce the marginal plants once they have been divided and repotted in fresh compost. The water level can then be brought up to the finished level.

In order to help with the rehabilitation of the fish, scatter small bunches of floating plants on the surface to reduce light levels and to provide as much surface cover for them as possible.

During a long cold spell, gases from decomposing plants must be allowed to escape from under the ice. A simple way to melt it is by placing a a bucket of hot water on the surface.

frozen ponds

Hard winters can mean the pond freezing up for weeks. Fish and plants can normally cope with this if the pond is deep enough to allow an area of unfrozen water at the bottom. Problems can occur when this condition persists for weeks and bubbles appear under the ice. These bubbles are likely to be methane from decomposing vegetation and normally released harmlessly to the atmosphere. When trapped for long periods under ice, the gas can be reabsorbed into the water and become toxic to fish.

REMEDIAL ACTION

If an electricity supply is near enough, small electric pond heaters are available to keep a small area of the pond clear of ice. Alternatively, place a pan of boiling water on the ice until the base of the pan melts enough of the ice to create a small clear circle of water. This process will need to be repeated daily or as often as it takes to keep the circle clear. Never try to break the ice with a hammer or any other implement. The strong vibrations caused by the banging will cause immense damage to the torpid fish.

If ice persists, make air holes through the ice layer to allow any methane gas to escape. In very severe prolonged freezing, the ice can expand on the surface and cause pressure on the walls of a concrete pond. This pressure can be absorbed by placing tennis balls or other absorbent objects on the surface when bad weather is forecast.

structural problems

PROBLEMS WITH CLAY-LINED PONDS

A well-maintained clay pond should last indefinitely provided every precaution is taken against root invasion from within and any surrounding trees. The main essential in maintenance is to ensure that it never dries out which could lead to permanent damage. Keep the pond topped up in dry weather and never allow the clay to become exposed.

If a pond is densely planted, there will be an inevitable build-up of decaying organic matter on the bottom which should be scraped out every five to seven years and the soil replaced. It is possible to introduce fish into clay-lined ponds, but avoid bottom-feeders such as carp, since they will be constantly stirring up the mud from the pool bottom.

If the pool level drops dramatically, a leak may have developed in the clay lining. This can sometimes be cured by sprinkling a liberal quantity of bentonite on the surface where any leak is suspected. It may be possible to trace the leak by pouring a vegetable dye such as fluorescin on the water surface to act as a tracer. If the trace is not clear, draining may be necessary and the dye will be particularly noticeable on the bottom where it has been drawn to the leak. Bentonite comes into its own as a repair material for clay ponds. Clear the soil over the suspect area and scatter the loose powdered material to a depth of 10–15cm (4–6in) over the clay before returning the topsoil and refilling.

LEAKING PONDS

Concrete shells

Old concrete ponds are the most likely candidates for leaks. Over the years, the concrete shell is subject to movement and strain in the surrounding soil, and hairline cracks appear, causing slow leaks. Inconsistency in the mixing of the original concrete plus the action of frost exaggerates the problems of concrete, which has been largely superseded by the use of flexible liners.

There are several proprietary concrete sealants that can be used once the pool is emptied and the offending cracks have been found. The area around the cracks should be thoroughly wire-brushed and a shallow 'V'-shaped channel cut along the crack with a cold chisel to provide a surface onto which the sealant can key. If the concrete has simply deteriorated and is flaking away, then the whole pond should be lined with a flexible liner.

Flexible liners

Leaks in flexible liners can be repaired more easily, once the leak is found in the liner. The level to which the pond drops is a clue to where the hole or tear can be found and it may not be necessary to drain the pond completely. Reduce the water level to just below the leak and apply a repair patch once the puncture area is dry and clean. There are repair kits available for most types of liner.

Fibreglass units

Preformed pools that are made of fibreglass are the least likely types of pond to leak, but in the event of a crack occurring it is better to repair this invisibly on the underside of the preform rather than to expose the repair in the pool. Suitable fibreglass repair kits are available from DIY motor body repair stores or suppliers.

BALLOONING FLEXIBLE LINERS

In certain soil conditions where there is a high water table, the pressure of water beneath the pond can be so great that it causes the liner to billow up into the pond water. This is most likely to happen in the winter when the water table is at its highest. If this happens frequently, you will need to consider putting in a drain from the side of the pond. If it only happens rarely after prolonged heavy rain, the solution is to place heavy stones or blocks on the bottom after the liner has returned to its normal level.

9 THE POND YEAR

Whatever the season, the surface of a pond will be a source of endless fascination, whether reflecting the winter sky in its still, mirror-like state or buzzing with insect life and bursting forth with colourful water lily flowers in the summer. Looking after the pond throughout the year will bring many rewards, from the greeting of pet fish in the spring as their appetites return with the warmer water, to enjoying autumnal tints in the leaves of marginal plantings that have been cosseted throughout the summer. In addition, regular maintenance will prevent problems occurring and ensure the best results.

spring

A well-designed informal pond where provision has been made for complementary planting beyond the water's edge will enjoy a peak of interest in the early flowers – a welcome sight after the subdued browns and straw-coloured textures of winter. The marsh marigolds (*Caltha palustris*) are amongst the first to appear. These yellow blooms make a perfect combination with many early bulbs that appreciate a wet soil in winter, particularly *Chionodoxa* and *Scilla* with their striking blue flowers. After the traditional snowdrops are finished, there is a waterside snowdrop, *Leucojum aestivum*, that grows much taller than the ordinary snowdrops, which makes a superb display against the dark backdrop of the clear water surface.

The pool water, which is at its clearest in early spring, allows the fresh green shoots of the sword-like iris leaves to be seen more easily as they break through the winter protection of the old brown foliage. As the days lengthen, the pond suddenly becomes alive with mating frogs, frantically splashing in the shallow margins before disappearing again as suddenly as they appeared. Frogspawn really marks the beginning of a new year in the water garden, as fish begin to show more life in the first of the warm days.

Many marginals start to show growth now, with the variegated manna or sweet grass (*Glyceria maxima* var. *variegata*) exhibiting lovely pink shades in the young striped leaves. The moist soil near the edge will see the primulas starting to flower, notably the drumhead flowers of *Primula denticulata* just above the soil surrounded by the looser pink flowers of *Primula rosea*. As the spring progresses, a great wealth of leaf colour and blooms appear, including the exotic yellow arum lily-like blooms of the skunk cabbages (*Lysichiton*) in advance of their truly enormous leaves.

General maintenance

- Remove pond heaters and replace with pumps to start up the filters.
- Check the pool water with a water-testing kit to see if it is acid, alkaline or ideally between the two extremes.
- If the pond has a high organic content that is decomposing (blackish water is symptomatic of this condition), carry out a partial water change of about one third of the pond's total volume.
- Clean off algae from paving and decking.

Left: Many moisture-loving plants suitable for surrounding pools flower in spring before the tree canopy increases and blocks out the sunlight.

- Repair timber bridges, decking or surrounds and treat with a non-toxic preservative.

Plant care

- Cut back the old brown leaves of the marginals if this was not done in the autumn. Cut the old leaves just above the young emerging green shoots, otherwise the new leaves will display the cut surfaces as they develop.
- Fork through surrounding borders to the pond and prune back shrubs such as dogwoods (*Cornus*) which are grown for the winter colour of their stems reflected in the water.
- Protect the young growth of specimen moisture-lovers and marginals, such as gunneras and rheums, against frost damage. Mound up old leaves over the emerging shoots, which can be held in place against the wind by chicken wire until frosts have passed.
- Drape fleece over frost-sensitive flowers such as skunk cabbages (*Lysichiton*) on frosty nights.
- Any of the slightly tender marginals, such as *Lobelia* and *Mimulus*, which have been protected through the winter by a heavy mulch, can now be lifted and divided. Softwood cuttings can also be taken from the young growth of these tender plants, which will then need the protection of a cold frame or greenhouse.
- The bright light conditions of spring can spark off algal growth in the pond when there is hardly any shade from water lily leaves. Scatter some floating plants on the surface to help reduce this light.
- Harden off seedlings or divisions that have been kept under cover for the winter.
- Plant moisture-loving perennials at the pool side.
- Consider any gaps in the planting with a view to making any necessary purchases in late spring.
- Mulch any surrounding moisture-loving beds to conserve moisture in them for the summer.

Fish care

- Start feeding fish again as the water temperature rises to about 10 degrees C (50 degrees F). Use a high-protein food and supplement with chopped worms.
- Examine the fish for any parasites or scale disorders that they may have picked up during the winter when they are in a weak condition.
- Herons are particularly hungry and active in the spring, so check that vulnerable ponds have protection.

summer

In early summer, the flowers and leaves of the globeflower (*Trollius*) in shades of orange and yellow create an eye-catching contrast to the bold upright foliage of the marginals. The creamy striped leaves of the yellow or flag iris (*Iris pseudacorus*'Variegata') make a particularly strong contribution to the planting before the leaves turn green in midsummer. Once the main spurt of early summer growth is over, the pond takes on a more restful appearance with the water lily flowers starting to open.

Late summer brings drama to the scene with the blooms of *Lobelia cardinalis* – the most strident red flowers to feature in the margins of the water. These blend well with another striking late summer plant, the pickerel weed (*Pontederia cordata*), whose spikes of beautiful pale blue flowers are carried above strong, glossy leaves that resemble warriors' shields.

General maintenance

- If the water level falls in hot sunny weather, it is not necessarily indicative of a leak in a small pond. It is more likely to be evaporation loss and should be replaced by frequent top-ups rather than leaving it too long and adding too much fresh water at one time. Use rainwater if at all possible since mains water is particularly enriched with mineral salts and chlorine during the summer months.
- Monitor the water chemistry

Left: The water surface becomes almost hidden as the marginals and moisture-lovers become fully grown. In contrast to the showy spring flowers, the light airy flowers of the *Alisma* (Water plantain) make a delicate contrast to the dark water.

with water-testing kits and treat accordingly.
- Keep the strainer clear on the front of pumps since these clog up easily and reduce the flow of the pump.

Plant care

- Divide any marginals or submerged plants as necessary. There are also several shallow water plants with floating leaves, such as the water fringe (*Nymphoides peltata*), which can be increased by simply separating the runners or plantlets from the parent plant.
- Introduce tender floating plants to the pool.
- Check water lily leaves for pests and blast off any with a powerful water spray.
- Keep removing the yellowing or dying water lily leaves and flowers by cutting the long stems well below the surface of the water.
- Apply special slow-release sachets of fertilizer to water lilies and any starved marginal plants.
- Cut back oxygenators if they grow too rampant and use the tips as softwood cuttings.
- Remove dead flowers to prevent seeding.
- Cut back hard any mildewed leaves and stems to encourage new growth before the autumn.
- Seed pods can be collected from aquatics at the end of their flowering period; the seed is best sown straight away for the majority of them, particularly the moisture-lovers such as primulas. Seed from the deep-water plants, such as water hawthorn (*Aponogeton distachyos*) and golden club (*Orontium aquaticum*), are sown thinly in seed pans or half pots of formed aquatic compost, which is kept constantly moist by standing in a shallow container of water.

Fish care

- Keep any fountains or waterfalls running throughout the night in hot weather, since any fish present will find this period of the night the most difficult in which to obtain enough oxygen.
- Remove blanketweed promptly before it forms thick mats which entangle small fish.
- Take care not to overfeed fish in the warm weather since there is more natural food available to them in the summer months.
- Introduce any new fish to the pool if required.

Left: Before the first frosts of autumn the tightly packed foliage of summer becomes thinner and flowers more sparse. A stray candelabra primula brightens an otherwise green scene before the autumn tints emerge.

autumn

Although autumn colours are associated with the tints of trees and shrubs, such as the Japanese maples (*Acer palmatum* cultivars and species), there are several waterside plants that produce subtle shades, such as the royal fern (*Osmunda regalis*). Throughout the summer its pale green leaves bring a refreshing coolness to the pool side, but as the nights turn colder, these turn to warm golden browns.

If the garden is big enough to accommodate a waterside tree, there can be few to beat the sight of a dawn redwood (*Metasequoia*) at this time of year. Although a conifer, it loses its leaves in the winter after they turn a soft yellow or orange colour. It is a narrow columnar tree that loves water near its roots and can be squeezed into places where broader trees would simply be impossible.

As the flowers of the water lilies wane, the scented white flowers of the water hawthorn (*Aponogeton distachyos*) enjoy a second flush of flowering. The tiny fronds of the floating fairy moss (*Azolla filiculoides*) turn an unusual shade of pink, supporting water-like globules of mercury that cling to the fronds. As the water temperature drops and the first frosts are felt, this delicate little plant sinks to the bottom to overwinter. Oxygenating plants, such as the parrot feather (*Myriophyllum aquaticum*), enjoy moving centre stage with the demise of the water lily leaves, and use the opportunity to scramble across the clear water, displaying their graceful feathery leaves against the dark surface.

General maintenance

- As bad weather begins, remove the pump if filters, waterfall or fountains are not going to be used for the winter. A floating pond heater can be connected to the socket released by the pump in readiness for the first severe spell of weather.
- Put a layer of netting (plastic mesh) over the pond surface, weighting down the edges with stones to keep it taut, to keep leaves away from the water surface.
- Continue removing blanketweed as necessary.

Plant care

- Protect frost-tender plants by providing a heavy organic mulch over their crowns.
- Remove any tender floating plants such as water hyacinth (*Eichhornia crassipes*) and water lettuce (*Pistia stratiotes*), and overwinter them indoors in a light, completely frost-free place. Put in shallow trays of compost that are kept constantly moist.
- The first frosts will blacken the leaves of large moisture-loving plants such as rheums and gunneras. Cut off the frosted leaves and drape them over the crown for protection over the winter.
- Cut back leaves of marginals around highly manicured formal ponds.
- Thin out and cut back submerged plants. Remove all the rotting and dying vegetation before it sinks to the bottom of the pool for the winter and decomposes.

Fish care

- If the weather is mild and water temperatures hold above 10°C (50°F), fish can still be fed but should have their diet changed to wheat-germ pellets. However, when it turns cold, stop feeding altogether.
- Provide protection for fish from predators by placing pieces of black plastic or terracotta pipe 60cm (2ft) long with a minimum diameter of 23–30cm (9–12in) on the bottom of the pond.

Above: Late winter and the first signs of spring growth appear with the iris, daffodils and marsh marigolds. The clear water surface will soon be alive with the calls of mating frogs and toads.

winter

The reflective qualities of water are at their height in this season, and perhaps surprisingly the pond can make its greatest impact of the year on an otherwise stark scene. Stems and silhouettes come into their own, particularly when reflected in the pool surface. In this respect, the dogwoods (*Cornus*) are a great asset at the pool side, especially the cultivar 'Westonbirt' whose vivid red stems stand out dramatically.

Despite the cold, birds make more use of the pond for bathing in winter and a small watercourse with rock pools which is kept running throughout the season will be highly valued. The wildlife in general will also appreciate the value of foliage being left to stand for the winter rather than the pond surrounds being cleared.

General maintenance

- If ice forms on the water surface for prolonged periods, make holes right through the ice layer to allow methane gas to escape.
- Protect shallow or raised pools that are more vulnerable to frost with a plastic sheet or boards in severe weather.
- Remove any snow lying on ice to allow light into the pool.
- Place floating objects on the water to absorb ice pressure.
- If freezing weather persists for unusually long periods, drop the level of the water by 5cm (2in), to create an insulating layer of air between the ice and the water.
- Periods of drying winds can occur in the winter, causing evaporation loss and a drop in level. Topping-up in winter, therefore, can be just as important as it is during the summer.
- Remove any dead leaves from the water.

Plant care

- Take in pygmy forms of water lily used in raised pools such as half barrels to over-winter in a frost-free place.
- Spray any neighbouring plum or cherry trees with a winter egg-killing spray in order to kill off any overwintering water lily aphids.
- Winter is a good time to look at the surrounding planting of the pond and make any adjustments in the shrub and tree selection.

Fish care

- If the pump is still operating a watercourse or biological filter, elevate the pump from the pond bottom so that it recirculates the colder water at upper levels of the pond. This allows the area of warmer water to be left undisturbed for the benefit of any fish.
- Leave fish well alone, and do not feed even in mild spells.

INDEX

Page references in *italic* refer to the illustrations

INDEX

ACKNOWLEDGEMENTS

Front cover top right: PHOTOS HORTICULTURAL.
Front cover top left: GARDEN PICTURE LIBRARY/Gil Hanly/Design: John Heywood, Auckland.
Front cover bottom: GARDEN PICTURE LIBRARY/Ron Sutherland/Design: Paul Fleming.
Back cover top right: GARDEN PICTURE LIBRARY/ Clive Nichols.
Back cover top left: GARDEN & WILDLIFE MATTERS.

GARDEN PICTURE LIBRARY/Mark Bolton 106/Brian Carter 19/Chelsea Flower Show 1996: Daily Mirror 'Garden of Rooms'/Designer: John Plummer 10/Ron Evans 78 Bottom /Vaughan Fleming 5 detail 7, 92 Bottom Left/Gil Hanly/Design: John Heywood, Auckland/Georgia Glynn-Smith 50/Sunniva Harte 95, 110 Top, 116 Bottom/Howard Rice 96, 97, 100//Ron Sutherland/Designer: Paul Fleming 2–3/Alec Scaresbrook 116 Top/J. S. Sira 80/J.S. Sira 66 Bottom Right/Ron Sutherland 71/Brigitte Thomas 33 Top Right/Juliette Wade 83 Top.
JOHN GLOVER 5 detail 6, 8 Bottom Right, 30 Top Left, 33 Top Left, 38 Bottom Left, 44, 52, 54 Top, 57 Top Right, 59, 63 Top Right, 66 Bottom Left, 72, 74 Top Left, 79, 85, 88, 92 Bottom Right, 93, 98, 105, 110 Centre/Chelsea 91 34/Chelsea 92/Hampstead Horticultural Society 56 left/Chelsea 93:Evening Standard/Designer Dan Pearson 109/Chelsea 95/Design: Mark Walker 5 detail 4/Chelsea 95/Designer: Mark Walker 55/Chelsea 96 46/Chelsea 96: Islington Borough Council 53/Chelsea 99/Designer: Paul Dyer 41/Designer: Pamela Woods 5 detail 1, 9/Designer: Susy Smith 18/Hampton Court 92 15, 32/Hampton Court 96:Sainsbury Garden 23 Top Right/Hampton Court 99/Designer: Guy Farthing 17.
HARPUR GARDEN LIBRARY 101/Jerry Harpur 22 Bottom Left/Jerry Harpur/Designer: Mark Laurence 5 detail 5, 66 Centre Right, 77/Jerry Harpur/Designer: Tom Hobbs, Seattle 48 Bottom/Marcus Harpur 37, 83 Bottom, 94.
ANDREW LAWSON 5 detail 2, 22–23, 26, 45, 86, 104, 107, 108 Bottom Left, 112/Designer: Andrew Card 61/Designer: Wendy Lauderdale 4–5, 5 detail 9, 118–119/Designer: Wendy Lauderdale 120–121, 122, 123/RHS Chelsea 1999/Glebe Cottage Plants:Carol Klein 81/Wollerton Old Hall, Shropshire 22 Bottom Right.
NATURE PHOTOGRAPHERS LTD/S. C. Bisserot 91.
NATURAL IMAGE/Bob Gibbons 68 Top Left, 108 Top/Mike Lane 75/Peter Wilson 89.
N.H.P.A./G. I. Bernard 103/Stephen Dalton 5 detail 8, 108 Bottom Right/Melvin Grey 90.
CLIVE NICHOLS PHOTOGRAPHY 5 detail 3, 38 Top, 38 Bottom Right, 64/Chelsea 1997/Designer: Geoff Whiten 22 Centre Right/Chelsea 94 12 Bottom/Chelsea 95 78 Top/Chelsea 97/Designer: Bunny Guinness 48 Top/Chelsea 99/Paul Dyer 12 Top/Designer: Richard Coward 68–69/Design: Stephan Woodhams 54 Bottom Left /Hampton Court 97:Garden & Security Lighting 65.
PETER ROBINSON 31 Top Left, 31 Top Centre Left, 31 Top Centre Right, 31 Centre Centre Right, 33 Centre Left, 43 Top Left, 43 Centre Left, 43 Bottom, 43 Bottom Centre Left, 63 Centre, 70, 73, 74 Top Right, 76, 84, 87, 92 Top, 102, 114.
DEREK ST ROMAINE 8 Top, 11, 13, 20, 24, 49, 60, 62, 67, 113, 115/Chelsea 1996/Designer: Julie Toll 66 Top Right/RHS Chelsea 1998: Lambeth Horticultural Society 8 Top Left/RHS Chelsea 1998: the Express Garden/Designer: Peter Hogan 6/RHS Chelsea 1999/Designed by Pershore College 54 Bottom Right/RHS Hampton Court 1997: Pet Plan Garden 5657/RHS Tatton Park Flower Show 1999/Designers: Paul Butler & John Roberts 39.